# OIL MARKETS IN TURMOIL

# OIL MARKETS IN TURMOIL
## An Economic Analysis

PHILIP K. VERLEGER, JR.

*Foreword by*
SENATOR BILL BRADLEY

*Contribution by*
MARSHALL THOMAS
Markets Editor
Petroleum Intelligence Weekly

BALLINGER PUBLISHING COMPANY
Cambridge, Massachusetts
A Subsidiary of Harper & Row, Publishers, Inc.

Copyright © 1982 by Ballinger Publishing Company. All rights reserved. No part of this publication may be reproduced, stored in a retrieval system, or transmitted in any form or by any means, electronic, mechanical, photocopy, recording or otherwise, without the prior written consent of the publisher.

International Standard Book Number: 0-88410-867-8

Library of Congress Catalog Card Number: 81-22810

Printed in the United States of America

**Library of Congress Cataloging in Publication Data**

Verleger, Philip K.
  Oil markets in turmoil.

  Includes index.
  1. Petroleum products—Prices. 2. Petroleum industry and trade. I. Title.
HD9560.4.V47      338.2'3      81-22810
ISBN 0-88410-867-8                AACR2

# CONTENTS

List of Figures ix

List of Tables xi

Foreword xvii

Preface xix

Introduction xxiii

*PART I*

    **Chapter 1**
    **The Economic Characteristics of an Interruption**    3
The Nature of a Disruption in Oil Supplies    5
The Relationship Between Price Determination, Demand, and
   Equilibrium Prices    21
Conclusion    26
A Note About Panic    26

## Chapter 2
### The Behavior of Oil Prices on Commodity Markets During Crises — 29

The Behavior of Prices on Commodity Markets During Disruptions — 30
Modeling the Movement of Spot Product Values — 39
Estimation of the Spot Product Value Equation — 46
Conclusion — 52

## Chapter 3
### The Relationships Between the Commodity Markets and Adjustments in Official OPEC Prices — 55

The Theory — 58
Empirical Test of the Theory — 61
Simulation Results and Conclusions on Price Determination — 65

## Chapter 4
### The Behavior of Consumer Prices — 67

The Relevance of Price Theory to the Oil Industry — 68
Regulation of Prices — 73
Behavior of U.S. Prices During the Iranian Disruption — 75
Behavior of Consumer Prices in Europe — 80
Conclusion — 87

## Chapter 5
### Inventories — 89

The Behavior of Inventories During the Crises of 1973 and 1979 — 90
The Reasons for Inventory Accumulation — 104
Application of the Theory of Inventory Accumulation to Oil — 111
Empirical Estimation of an Inventory Model — 114
Conclusion — 121

## Chapter 6
### Modeling the Behavior of the Market During Disruptions ... 123

Consumer Demand, Shortages, and Changes in Expected Supply ... 125
The Behavior of the Oil Market ... 127
The Effect of Variations in the Parameters of the Oil Model ... 134
The Effect of Alternative Rates of OPEC Price Adjustments on Market Stability ... 135
The Effect of Alternative Means of Setting Consumer Prices on Market Stability ... 137
Conclusions ... 145
Postscript ... 147

# PART II

## Chapter 7
### The Effectiveness of Price Controls During a Disruption ... 155

The Form of Price Controls ... 155
Quantification of the Effects of Price Controls ... 157
Conclusion ... 165

## Chapter 8
### Tax and Tariff Policies for Coping with Disruptions ... 167

Possible Forms of Disruption Taxes and Tariffs ... 168
Quantitative Assessment of Tax Options ... 176
The Problem of Retaliation ... 193
Conclusion ... 196

## Chapter 9
### Policies to Encourage Building Inventories Before Disruptions ... 199

Incentives to Build Supplier Inventories ... 200
Increasing the Potential Profit from Holding Stocks ... 207

Incentives to Build Consumer Inventories 215
Building Publicly Owned Stocks 217

**Chapter 10
Policies to Reduce Consumption During
a Disruption** 221

Emergency Measures for Reducing Consumption 222
The Effects of Implementing Emergency Conservation
 Measures as a Means of Moderating the Effect of
 Disruptions 239
Conclusions 244

**Chapter 11
Policies for Managing Disruptions** 245

Elements of an Effective Policy 246
The International Energy Agency as a Policy Instrument 255
Conclusion 260

**Appendix
The ABCs of Measuring Oil Market Price Trends**
*— Marshall Thomas* 263

References 275

Index 281

About the Author 291

# LIST OF FIGURES

| | | |
|---|---|---|
| 1–1 | Static World Supply of Crude Oil | 7 |
| 1–1a | Static World Supply of Crude Oil | 8 |
| 1–2 | Static Depiction of Reduction in Supply | 9 |
| 1–3 | Change in Static Supply Curve for World Oil Illustrating Response to Saudi Sale of Incremental Production During Iranian Shutdown | 11 |
| 1–4 | Effect of a Shift in Supply Is to Raise Price From $P_o$ to $P_1$ if Price Change Is Simultaneous with Shift in Supply | 14 |
| 1–5 | Contract (or Posted) Price of Crude Follows the Spot Value | 17 |
| 1–6 | Effect on the Short-Run Equilibrium Price of a Failure by OPEC to Raise Prices Is to Drive that Price Higher | 22 |
| 1–7 | Effect on Short-Run Equilibrium Price of Consumer Hoarding Response Is to Drive Prices Even Higher | 24 |
| 1–8 | Equilibrium Is Restored when OPEC Raises Floor Price from $P_o$ to $P^*$ | 25 |
| 2–1 | Market Response to Supply Disruption in Three Past Crises | 32 |
| 2–2a | Identification of the Demand Curve During a Disruption | 44 |
| 2–2b | The Price Coefficient of the Demand Curve Cannot Be Identified During Periods of Stable Supply at Constant Prices | 44 |

## LIST OF FIGURES

| | | |
|---|---|---|
| 4-1 | Effect of a Disruption on Product Market Equilibrium | 69 |
| 4-2 | Comparison of Official Price of African Light, Spot Value of African Light, and Normalized Value of Consumer Product Prices in the United States | 79 |
| 4-3 | Comparison of Official Price of African Light, Spot Value of African Light, and Inland Refiner Realizations in West Germany | 86 |
| 5-1 | Comparison of Spot Market Value of Products from a Barrel of Crude on Gulf Coast Markets with the Post-entitlement Refiner Acquisition Cost of Crude | 112 |
| 5-2 | Comparison of Petroleum Inventories (Not Seasonally Adjusted) with the Profit on Products from a Barrel of Oil Sold at *Platt's* Gulf Coast Spot Prices | 113 |

# LIST OF TABLES

| | | |
|---|---|---|
| 1-1 | Prices of Principal Petroleum Products on Rotterdam Market | 13 |
| 1-2 | Spot Product Values (Netbacks) and Official Prices of African Light Crude, 1978 to 1980 | 16 |
| 1-3 | Actions by Members of OPEC that Increased Prices During 1979 | 18 |
| 2-1 | Comparison of Salient Characteristics of Six Episodes | 34 |
| 2-2 | Comparison of Percentage of Change in Price and Output During Six Episodes | 36 |
| 2-3 | Normative Observations on Six Episodes of Sudden Change in Oil Markets | 39 |
| 2-4 | Monthly Pattern in World Crude Oil Production | 48 |
| 2-5 | Results from the Econometric Estimation of the Adjustment Equation | 50 |
| 2-6 | Results from the Econometric Estimation of the Spot Value of Adjustment Equation Estimated in First Difference Form | 51 |
| 3-1 | Netbacks on Various Types of Crude Oil During the First Months of 1981 as Published in *Platt's* | 57 |
| 3-2 | Estimates of the Parameters for the Unconstrained Price Equation | 62 |
| 3-3 | Estimates of the Parameters for the Constrained Price Equation | 63 |

| | | |
|---|---|---|
| 3-4 | Test of Hypothesis $H_1$, that $W \sum_{k=1}^{\infty} \alpha^i = 1$, versus $H_o$, that $W \sum_{k=1}^{\infty} \alpha^i \neq 1$ | 64 |
| 3-5 | Mean Simulation Errors for Estimates of the Constrained Price Equation | 65 |
| 4-1 | Comparison of Spot Product Values and Spot Crude Oil Prices of Mideast Light and African Light, 1978-80 | 72 |
| 4-2 | Cumulative Increase in Consumer Prices of Various Petroleum Products Since June 1978 in the United States as Compared to the Increase in the Cost of Crude | 76 |
| 4-3 | Cumulative Increase in U.S. Weighted Average Product Prices from June 1978 as Compared to Crude Oil Price Increases | 77 |
| 4-4 | Cumulative Increase from October 1973 in Consumer Prices of Gasoline and Diesel in Various Countries During the Arab Embargo as Compared to the Price of Contract Price of Crude Oil and the Spot Value of Crude Oil | 81 |
| 4-5 | Cumulative Increase in Consumer Gasoline Prices from June 1978 in Various European Countries, 1978-80, as Compared to Increase in the Contract Price of Crude Oil and the Spot Value of Crude Oil | 83 |
| 4-6 | Cumulative Increase in Consumer Prices of Diesel Fuel from June 1978 in Various European Countries as Compared to Increase in the Contract Price of Crude Oil and the Spot Value of Crude Oil | 84 |
| 4-7 | Cumulative Increases in European Refinery Prices from 1978:3 as Compared to Increases in the Cost of Crude and the Spot Value of Crude Oil, 1978:4 to 1980:2 | 85 |
| 4-8 | Cumulative Increase from 1978:3 in Weighted Average of European Refinery Product Prices as Compared to Increases in African Light Crude Prices, 1978:4 to 1980:2 | 87 |
| 5-1 | Sources of Refined Products Supplied to the U.S. Market and Stocks of Crude Oil and Products, 1972:4 and 1973:4 | 92 |
| 5-2 | Sources of Refined Products Supplied to the U.S. Markets and Stocks of Crude Oil and Products, 1973:1 and 1974:1 | 94 |

| | | |
|---|---|---|
| 5-3 | Sources of Refined Products Supplied to the U.S. Market and Stocks of Crude Oil and Products, 1973:2 and 1974:2 | 95 |
| 5-4 | Comparison of Quarter-to-Quarter Changes in Supply and Consumption in the United States, Japan, and Europe in 1973 and 1974 | 96 |
| 5-5 | Sources of Petroleum Products Supplied to the U.S. Market, 1978:1 and 1979:1 | 98 |
| 5-6 | Sources of Refined Products Supplied to the U.S. Market and Stocks of Crude Oil and Products, 1978:2 and 1979:2 | 99 |
| 5-7 | Sources of Refined Petroleum Products Supplied to the European Market, 1979:1 versus 1978:1 and 1979:2 versus 1978:2 | 102 |
| 5-8 | Sources of Refined Products Supplied to the Japanese Market, 1978:1 versus 1979:1 and 1978:2 versus 1979:2 | 103 |
| 5-9 | Comparison of Change in Available Supply and Change in Stocks Between the United States and Europe, 1979:1 and 1979:2 | 104 |
| 5-10 | Days of Supply of Oil Held by OECD Countries | 108 |
| 5-11 | Indicators of the Increase in U.S. Value of Inventories in Manufacturing and Petroleum, Quarterly from 1978 to 1980 | 109 |
| 5-12 | Inventory Demand Equations Estimated by Ordinary Least Squares | 116 |
| 5-13 | Inventory Demand Equations Estimated by Instrumental Variables | 118 |
| 5-14 | Comparison of Actual and Predicted Rates of Acquisition of Stocks of Crude Oil, 1979 and 1980 | 121 |
| 5-15 | Comparison of Actual and Predicted Rates of Acquisition of Stocks of Products, 1979 and 1980 | 121 |
| 6-1 | Summary of Model | 124 |
| 6-2 | Published Estimates of Short- and Long-Run Price Elasticities of Demand for Petroleum Products | 126 |
| 6-3 | Parameters Used in Model of Oil Markets | 129 |
| 6-4 | Initial Simulation of Oil Model | 132 |
| 6-5 | Number of Months Required by OPEC to Capture 90 percent of any Increase in Spot Values | 136 |
| 6-6 | Spot Values of Crude Oil Under Different OPEC Pricing Regimes | 136 |

| | | |
|---|---|---|
| 6-7 | Impact of Substituting Consumer Prices Based upon Spot Product Values for Consumer Prices Based upon Official Crude Prices | 138 |
| 6-8 | Impact of Basing Consumer Prices on Spot Values in Lieu of Official Crude Prices, Assuming that Spot Market Pricing Negates Incentive to Build Stocks | 140 |
| 6-9 | Impact on Spot Values of Initial Stock Levels that Are Above or Below Equilibrium Levels at the Time of a Disruption | 141 |
| 6-10 | Comparison of Spot Values Under Alternative Inventory Acquisition Assumptions | 142 |
| 6-11 | Comparison of Spot Values of Crude Oil Under Different Parameters of the Demand Function | 144 |
| 6-12 | Model Performance in Assessing Movement of Spot Values and Official Crude Prices During the Iranian Crisis November 1978 to June 1979 | 149 |
| 6-13 | Comparison by Actual and Predicted Spot and Official Prices During the Iran/Iraq War | 150 |
| 6-14 | Comparison of Predicted Spot Price Assuming that Inventories Are High at the Start of the Iran/Iraq War with the Case Where Inventories Were Low | 150 |
| 7-1 | Predicted Spot Crude Values Under Three Different Shortage Scenarios with Different Price Control Regimes | 161 |
| 7-2 | Predicted Official Crude Prices Under Three Different Shortage Scenarios with Different Price Control Regimes | 162 |
| 8-1 | Assessment of Effectiveness of Various Tax Proposals in Dealing with a Disruption | 173 |
| 8-2 | Comparison of the Effect of Three Tax Programs on Changes in Spot Values Under Two Disruptions | 178 |
| 8-3 | Levels of Variable Tariff Required to Meet a 4 Million Barrel a Day Disruption and Prevent Further Increases in Spot Values | 184 |
| 8-4 | Levels of Tariffs Used to Restore Spot Values to Predisruption Levels | 185 |
| 9-1 | Comparison of Year-to-Year Changes in Official Sales Prices of African Light Crude (January prices) | 201 |
| 9-2 | Calculation of Break-Even Prices for Crude Oil Held More Than One Year (Based on African Light) | 203 |

| | | |
|---|---|---|
| 9-3 | Estimated Annual Cost of Holding 90 Days of Sales in Mandatory Stocks for Eleven U.S. Oil Companies | 214 |
| 9-4 | Oil Stockpiling Programs of Eighteen IEA Countries | 218 |
| 10-1 | Estimates of Potential Conservation by Building Temperature Controls | 225 |
| 10-2 | Estimates of Oil Savings Available by Fuel Switching | 228 |
| 10-3 | Estimates of Oil Savings Through Power Wheeling | 229 |
| 10-4 | Estimates of Reductions in Personal Fuel Use by Three Different Studies | 234 |
| 10-5 | Estimates of Emergency Conservation Potential | 238 |
| 10-6 | Effects of Conservation Measures on Increases in Spot Crude Values with an Interruption in Supply | 241 |
| 10-7 | Size of Emergency Conservation Measures Required to Offset a Disruption of 4 Million Barrels a Day | 242 |
| 10-8 | Effect of Joint Tariff Conservation Measures as Mechanisms for Offsetting a 4 Million Barrel a Day Loss in Supply | 243 |
| 11-1 | Supply and ERDO Calculations for Two Disruptions | 257 |

# FOREWORD

The most pressing aspect of energy policy today is the preparation for the disruptions in our oil supply that are virtually inevitable over the next decade. This aspect of energy policy has been neglected, even though three relatively small disruptions—the Arab embargo in 1973, Iran in 1979, and Iran/Iraq war in 1980—have had terrible consequences for our standard of living and for our economic growth. Constant turmoil in the Middle East underscores the need to prepare. The best way to prepare for oil supply disruptions is to stockpile oil in advance. My longstanding advocacy for the strategic petroleum reserve reflects this concern about preparedness. But what happens if a disruption occurs before our reserves are sufficient? The United States will face two unappealing choices: either rapidly rising oil prices or government price controls, oil allocations, and even rationing.

After ten years of controls, can it really be said that they protected the nation as a whole or the poor in particular? Have price controls stopped price increases?

Price controls cannot sufficiently protect the poor from an international market in which only part of the supply is controlled. Price controls encourage oil consumption during a disruption and discourage production of alternatives for imported oil. Price controls send false and damaging signals to the economy that oil is not scarce, that conservation is not needed, and that alternative sources of energy are unimportant. Finally, even the threat of price controls reduces most individual incentives to prepare for disruptions.

In the future, we must choose between a regulatory approach to oil supply disruptions and a market approach. Clearly, neither option will work perfectly. The market approach will inflict short-term pain in the form of higher prices and income transfers within the country; the regulatory approach will inflict pain by creating physical shortages, higher oil prices, and excessive economic costs.

The relative magnitudes of the costs associated with each approach have not been carefully analyzed in the past. Indeed, the United States has entered each of the three supply disruptions of the last decade with price controls in place. We never before have had to decide between the regulatory and market approaches. Before the next disruption we will have the opportunity to choose. This study will assist us in that decision.

Dr. Verleger provides a quantitative description of the behavior of oil markets during disruptions and provides a critical assessment of alternative policies designed to address the problems of disrupted oil markets. I believe this analysis demonstrates conclusively that the regulatory approach will impose far greater costs than a market approach. Further, Dr. Verleger demonstrates that certain types of tariff policies might be used to strengthen the market solution and reduce the impact of an emergency.

Three times in the last eight years we have seen what happens when Washington tries to control prices and allocate oil supplies. We know our supplies can be interrupted again at any time. This study suggests options for government responses that can minimize, rather than exacerbate, the effects of an oil supply disruption.

Washington, D.C.  **Senator Bill Bradley**
October 1, 1981  Member, Senate Committee on
Energy and Natural Resources

# PREFACE

This book began as a study of the effects of federal price controls on the development of the oil industry during the 1970s. However, as the work progressed it became apparent that the original subject of petroleum price and allocation controls was far too broad to be covered thoroughly in one volume and that the issue would become moot before the book could be published. The scope of the book therefore was narrowed to focus on the behavior of oil markets during disruptions. Although some have viewed this behavior as resulting entirely from economic regulation, this study concludes that causes are complex.

In preparing this study, I received the assistance of many individuals whom I would like to thank individually and collectively. Chris Jones, Richard Eaton, and Peter Mathieu provided organizational support at Yale, without which this book could not have been written. In particular, Chris Jones arranged financial support for a two-day conference on "New Strategies for Managing U.S. Petroleum Disruptions" held in November 1980, which generated the ideas expressed in this volume. Peter Mathieu provided the excellent logistical support that enabled the conference to run smoothly, and Richard Eaton arranged financial support for both the conference and the preparation of this volume.

I would also like to thank Geoffrey Hazard, Acting Dean of the School of Organization and Management at Yale, who was willing to give his full support to the conference. Finally, I would like to thank the Carthage Foundation and the Program on Business/Government Relations at Yale for financial support.

I would like to thank the Editors of *Petroleum Intelligence Weekly*, and particularly Georgia Macriss, for allowing me to republish the *PIW* supplement of February 2, 1981, which describes the value of the different petroleum markets. This supplement clearly explains the differences that most economists have been unwilling or unable to understand in the past. I would also like to thank the editors of the *Review of Economics and Statistics* for permitting me to reproduce in part my article on the behavior of OPEC prices (in Chapter 3).

Many colleagues at other universities, in government, and at various oil companies provided helpful criticism of various drafts of this work and of the preliminary papers that formed the basis of the book. I am particularly grateful to Jim Griffen of the University of Houston, Bill Hogan of the Kennedy School at Harvard, and Robert Milbrath, a retired Vice President of Exxon; I would aslo like to thank Bill Taylor of Senator William Bradley's staff for reading many of the chapters and suggesting major revisions. Arnold Moore and Mike Canes of the American Petroleum Institute prepared detailed comments and suggestions on the early chapters of the book, and Tony Finizza, now of Arco, located several errors and suggested a very useful restructuring of part of the book.

I would also like to thank David Munroe, of the President's Council of Economic Advisers, Edward Erickson of North Carolina State University, David Wood, Henry Jacoby, Tom Neff, and Allen Jacobs of M.I.T, Alvin Alm and David Deese of Harvard's Kennedy School, and Martin Shubik of Yale for their assistance at various points in this endeavor. Pauli Colburn and Jim Bressler of Booz, Allen & Hamilton also helped in the proof-reading of the page proofs.

Special thanks are owed to two individuals for giving unselfish advice, counsel, and support. Linda Scotten served as secretary, research assistant, editor, and advisor for the two years I was at Yale. In this role she revised, typed, and challenged an unending stream of unintelligible drafts. Thanks to her diligence, this book is written in understandable English. Her extraordinary efforts are gratefully and humbly acknowledged.

Morris Adelman provided extraordinary support as mentor, colleague, collaborator, and friend. He provided the road map, guidance, and encouragement necessary to complete the work. Indeed, he has forgotten more about the economics of the oil market than most of us, particularly those of us working in New Haven rather than Cambridge, will ever know. The credit for any new ideas to be found in this volume belongs to Morris.

I would also like to thank the staff at Ballinger for their assistance in preparing this book. Special thanks are owed to Steven Cramer and Rosemary Winfield for their detailed editorial assistance. I am also indebted to Carol Franco for offering support and encouragement throughout this project.

My greatest debt of all, however, is owed to Margaret and Katy, who encouraged me to write this book and who tolerated my prolonged absences while it was being written.

Guilford, Connecticut  
August 1981

**Philip K. Verleger, Jr.**

# INTRODUCTION

The disruptions in world oil markets in the 1970s raise a series of economic issues concerning the nature of disruptions and the most effective methods of dealing with them. Of particular interest is the response of consumers, oil companies, and oil exporting nations. According to the conventional wisdom, consumers are innocent victims afflicted by shortages in the short run and higher prices in the long run, while oil companies and oil exporting nations are greedy extortionists who seize the opportunity created by a disruption to double or triple the price of oil.

The governments of consuming countries have formulated emergency policies based on this belief and on the belief that the shortage created by the disruption, if not met by government action, would wreak widespread economic chaos: important economic activity would slow or stop; prices would increase; and some consumers would suffer great hardship while others realized large windfalls. Several types of governmental actions have been introduced in past crises to prevent these consequences, including imposing price controls to reduce the hardship inflicted on consumers and deny windfalls to producers; using emergency measures to cut consumption of oil and possibly finance the use of other energy sources; and using inventories to make up any remaining gap between supply and demand.

During the 1973-74 Arab embargo, price controls were imposed in the United States and other major consuming countries; end use allocations of oil were established to assure that users whose activities were vital to the economy received adequate supplies; and emergency conservation measures were instituted. Despite these actions, the price of oil tripled from less than $4.00 to more than $10.00 a barrel in a period of six months.

These same policies were tried again during the 1979 interruption in supply caused by the collapse of the Iranian government. Voluntary and mandatory price controls were set; allocations to priority users were imposed; and formal agreements to cut consumption were established between the member countries of the International Energy Agency. Nevertheless, oil prices rose again, from $12.50 to $25.00 a barrel in twelve months.

Despite the admitted failure of these policies to stop price increases in 1973 and 1979, they have not been greatly modified to prepare for future interruptions. While price controls and allocations probably would be less appealing in the future, the principal means of meeting a crisis would remain the use of emergency conservation measures, possibly combined with attempts to draw down inventories. This study concludes that such policies would again fail to prevent prices from increasing.

Prices can be expected to increase greatly during the next supply interruption. Because little effort has been made to understand the characteristics and behavior of oil markets during disruptions, the response to the interruption probably will be too slow and in the wrong direction. In particular, a lack of understanding of the behavior of spot markets, of consumer demand, and of the determinants of inventory demand have led to a belief that a disruption can be met by matching the loss in supply with an equal cut in consumption (or with reduced consumption augmented by stock drawdown). This would be true only if the cut in consumption could be made simultaneously with the cut in supply, which is extremely unlikely for many reasons: emergency measures take time to implement; the size of a shortage is neither static nor readily measurable; and consumers (voters) generally resist and thus dilute the effectiveness of measures that impinge on their established routines. If the cut in consumption is not made simultaneously with the cut in supply, the resulting shortage starts a process that quickly develops a momentum of its own, ultimately causing adjustments in prices far in excess of those required to rebalance supply and demand under the new supply conditions.

This study describes the forces that push prices beyond the equilibrium levels by specifying the behavior of the oil market during disruptions. The word that most closely describes the behavior of consumers, producers, and refiners is "lethargic," not "greedy." The market participants (consumers, companies, governments, producers) adjust slowly to changes in market conditions and this lethargic adjustment transforms a minor incident into a major crisis and a major crisis into a catastrophe. Chapters 1 through 6 provide a description of this adjustment process.

Once this characteristic of market adjustment is recognized, rational and effective measures for meeting a crisis can be developed. Furthermore, the perversity of other, irrational policies, such as price controls, also can be assessed. Part II of this study (Chapters 7 through 11) offers assessments of various policy alternatives.

Throughout the study, the primary focus is on the spot market for petroleum products—commonly referred to as the Rotterdam market. This market is a barometer of conditions on world oil markets, rising during periods of shortage and falling during periods of surplus. It is in this market that unsatisfied demand and uncommitted supply meet. The exchange sets the spot price, and the spot price, in turn, determines the world price.

This market also offers a means of measuring the effectiveness of various measures imposed at times of disruptions. Successful measures stop the process of price increases. Here there is an exact analogy to actions taken by central bankers in defending currency. A successful defense of currency stops foreign exchange rates from falling, just as successful energy measures stop prices from increasing.

The movement in spot prices during disruptions is traced to the size of the shortage. The greater the shortage, the larger the price increase that accompanies it. The analysis also shows that the increase in spot prices continues as long as the shortage persists. Thus, delays in response to a shortage drive prices higher. This characteristic has been recognized by many and has caused some to argue that controls should be placed on the spot market. Proposals of this sort are viewed here as analogous to hanging the messenger for delivering news of defeat.

The analysis presented in Part I finds that the sources of problems in oil markets during disruptions are not those commonly listed by critics of the oil industry. For instance, OPEC is more guilty of lethargy than greed. It uses the spot market for petroleum products to set crude prices but tends to follow that market very slowly. This slow adjustment creates an incentive for those having access to low-priced OPEC crude to profit from the temporary difference between the price set by exporting countries and spot market prices. This characteristic tends to prolong the disruption and increase the magnitude of the price increase. *Thus, one conclusion of the study is that the problems of disruptions could be dealt with more easily if OPEC would raise prices more quickly.*

*A second finding of this study is that oil companies tend to raise consumer prices too slowly during disruptions.* Increases in consumer prices tend to follow increases in OPEC prices. Thus, to the extent that OPEC is culpable for its lethargy, oil companies are culpable for following OPEC rather than the spot market in setting prices. Some or all of this guilt must be shared by governmental policies that have prevented oil companies from raising prices. Even where no government restrictions exist, company officials can be excused for not raising prices out of a justifiable fear of the wrath of government officials.

*A third finding of this study is that the consumer contributes to the process of increasing prices.* This contribution occurs through the slow consumer response to higher prices, demonstrated by the difference between short- and long-run price elasticities of demand. By responding slowly to changes in prices, consumers create conditions that tend to force prices even higher. Prices are driven still higher when price controls or other considerations delay the increase in consumer prices.

Slow adjustment in prices and unwillingness or inability to increase consumer prices also contribute to the ultimate panic that accompanies a disruption by making it profitable to increase inventories during a disruption. It becomes possible to buy at low prices during the initial days of a disruption and then sell at much higher prices by the end. Quite naturally, this causes inventory demand to increase, which increases the size of the crisis.

In short Part I shows that oil producers, oil consumers, oil companies, and governments of consuming countries all play a part in making an interruption a catastrophe instead of a minor inconvenience. The catastrophe is avoidable, however, by prompt and effective governmental action. In Part II, several alternative measures are assessed. The following recommendations emerge from this assessment.

### Recommendation One: Let prices rise quickly.

Interruptions in the supply of oil can be dealt with most easily by aggressively raising prices, because higher prices start the conservation process required to return supply and demand to equilibrium. Furthermore, a quick increase in prices will prevent oil prices from "ratcheting" to even higher levels by the end of the crisis. Since OPEC countries have been reluctant to raise prices quickly in the past, the governments and oil companies in consuming countries should be willing to initiate the price increase.

### Recommendation Two: Raise prices by imposing a large tariff on the importation of oil.

A tariff on imported oil will raise consumer prices and induce quick conservation. It will also reward those who have built speculative inventories of crude oil and product by allowing the holders of these stocks to sell them for higher prices (inclusive of tariff) without sharing the windfall with the government. This feature will increase stock holdings. Furthermore, release of the increased stocks will tend to dampen the increase in spot prices and thus benefit consumers through lower long-run oil prices. A tariff also will reward those who develop capacity to produce incremental supplies of oil or other fuels, to the extent that these gains are not captured by taxes.

### Recommendation Three: Adopt programs to encourage the development of greater private stockpiles.

The impact of a disruption on prices depends on the level of private stockpiles: the greater the stockpiles, the smaller the impact on prices. Thus, the adoption of measures that either reduce the cost of holding stocks (such as tax subsidies) or increase the expected profits from holding stocks in the event of a disruption should be considered.

> **Recommendation Four: Consuming countries should use a graduated tariff on oil imports to control the rate of inventory drawdown and defend any oil price they select.**

In the long run, consumers must cut consumption to match any reduction in supply. The way in which the cut is managed will determine the rate at which oil prices rise. This study advocates the use of a graduated tariff that starts at a high value and then quickly is reduced over several months. A graduated tariff would:

- Cause a maximum reduction in consumption by overcoming the inertia that characterizes consumer demand;
- Cause a maximum rate of stock sales in the early months of a disruption by offering the maximum profit to speculators who sell early;
- Minimize the draw on strategic stocks by using the cut in consumption and sales from private stocks to partially relieve the shortage; and
- Encourage consumers to postpone consumption by offering the promise of declining rather than rising prices.

The materials presented in Chapter 8 suggest that a disruption tariff can be used to defend any oil price in the same sense that a central bank defends the value of a currency. They also suggest that the volume of stocks used during the early months of the crisis will depend upon the size of the tariff and the price that is defended.

> **Recommendation Five: Establish mechanisms for recycling the receipts of the tariff.**

Many studies have shown that an interruption in oil supplies would have serious consequences for the U.S. economy due to the increase in the price of oil and the transfer of wealth from the U.S. to producing countries. The imposition of a disruption tariff would block this transfer of wealth, but the macroeconomic impact of a disruption can be minimized only if mechanisms that allow the instantaneous rebate of tariff proceeds are established before the disruption.

### Recommendation Six: Do not treat refiner and consumer taxes as substitutes for a tariff.

Suggestions that taxes on gasoline or refiner output be imposed in lieu of a tariff negate the potential benefit of speculation. Specifically, the imposition of a tax on a product denies the speculator the windfall from the liquidation of stocks. If standby taxes are enacted in lieu of a tariff, speculators will hold fewer stocks and tend to husband them longer during a disruption. Both actions will magnify the price increase that accompanies a disruption.

### Recommendation Seven: Make access to publicly owned stockpiles easy and quick.

One of the most significant responses to the Arab oil embargo was the decision to develop strategic oil stockpiles in many consuming countries. Given this study's findings about the effect of stockpiles in meeting disruptions, this decision was clearly correct. However, these stockpiles are of no use unless the oil is accessible quickly at the start of a disruption. Some mechanism must be established to enable consumers (not governments) to control the timing and the rate of withdrawal. This is particularly important since publicly owned stockpiles tend to reduce the size of privately owned stocks.

### Recommendation Eight: Impose emergency conservation measures quickly.

Emergency conservation measures represent a useful element in any program designed to meet a disruption. To be effective, however, the programs must be imposed quickly and must make real reductions in consumption. Unfortunately, one finding of this study is that the conservation measures that appeared promising in 1973 or 1979 either did not work or worked so well that they have now been fully incorporated into daily life and thus offer little further potential for conservation. In a future disruption, the only measures guaranteed to work are the politically unpopular ones, such as banning weekend driving.

### Recommendation Nine: Do not impose controls and allocations.

The standard response to a disruption has been to impose price and allocation controls. No action could be more detrimental to the long-run interest of consumers and consuming countries, because controls delay adjustment and drive prices higher. Furthermore, because allocations provide special treatment for some companies or some consumers, they remove the incentive for these companies and consumers to build precautionary stocks. This reduces the world's stockholdings and increases the probable price effect of any disruption.

### Recommendation Ten: International cooperation is a good idea but not essential to meeting a disruption. Further, the present structure of international cooperation as set up in the International Energy Agency is counterproductive.

The analysis developed in this study assumes that all developed countries impose uniform measures to meet a disruption. All countries do not have to respond in the same fashion, but their response must be quick, must calm the spot market, and must reduce consumption to the new level of supply before stocks are exhausted.

While international cooperation is helpful and perhaps even essential in meeting a disruption, coordination through the International Energy Agency (IEA) as it presently is constituted probably will prove to be a disastrous mistake for two reasons. First, the IEA as an international consultative body cannot move quickly. While diplomats and bureaucrats from the member states meet to discuss an ongoing disruption, consuming countries will be reluctant to take independant action for fear of losing something at the bargaining table (as happened in 1979). This inaction causes prices to spiral ever upward and panic to develop. This effect could be countered by giving the IEA authority to take unilateral action at the start of a disruption. It is unlikely, however, that the member states would assign it such authority.

The second problem with the present IEA arrangement is its obsession with quantities and the allocation of those quantities. Although all IEA emergency plans address shortages of varying magnitudes, interruptions are not primarily a problem of quantity but of price. It is the increase in price and the transfer of wealth, not the shortages, that create the great economic dislocations. Rational energy policy can and should be designed to achieve the sudden reductions in consumption necessary to meet a loss in supply without allowing prices to rise indefinitely. The thesis of this study is that these objectives can be met only if energy policy shifts its focus from quantity to the real villain—price.

# PART I

# 1 THE ECONOMIC CHARACTERISTICS OF AN INTERRUPTION

> Western oil companies are expert at accumulating their oil stocks in the most disruptive way. They buy when shortage threatens, thereby raising prices and increasing the shortage. They run down these stocks when the panic is over and the cost of holding them becomes a burden. Come the glut, the companies increase it, plunge real prices down, and strengthen OPEC's determination to cash in next time. Come the shortage, they dash for excess stocks (and for large expected profits from stock appreciation).
>
> *The Economist*, (September 27, 1980): 13.

This statement from *The Economist*, published three days after hostilities broke out between Iran and Iraq, succinctly summarizes the characteristics and shibboleths attending the issue of a disruption in oil supplies: The market liquidates inventories during periods of market stability (glut) and builds during periods of disruption (shortage), and OPEC countries raise prices in response to this pattern. This is not, however, a complete picture of the behavior of oil markets during crises, because it fails to identify the specific linkage between the shortage and oil company behavior. Furthermore, it fails to explain the mechanisms by which members of the Organization of Petroleum Exporting Countries (OPEC) translate shortages into price increases.

The purpose of the first half of this volume is to complete the description of the behavior of oil markets during disruptions by identifying the various structural elements and quantifying their interrelationships. Oil company stock acquisition policies do conform to the description posited by the editors of *The Economist*, but they are not the only causal factor in the train of events that follows the start of a disruption. Three other major behavioral characteristics can be identified, and each plays an important role: the process by which OPEC adjusts prices; the manner in which consumer prices of petroleum products are established; and the way in which prices on spot or commodity markets change during a disruption.

A loss in supply initially triggers an increase in spot prices. Since this increase is not translated into an immediate increase in either official crude prices or consumer prices, an excess demand for oil is created. This puts additional pressure on world markets, resulting in further increases in prices on commodity markets and causing producers of crude (members of OPEC) to make adjustments in their price postings. Because the adjustment has been slow in the past, however, we find there has been a tendency for official crude prices (postings) to rise beyond the postdisruption equilibrium level. In addition, since some oil suppliers have been willing to reduce output to maintain this new price level, unnecessary long-term damage to the economies of consuming countries has resulted.

The purpose of this chapter is to describe the theoretical dynamics of a disruption, beginning with the economic nature of a disruption in supply. This shift in supply induces simultaneous, but perhaps transitory, price increases on world commodity markets, particularly on the Rotterdam market, and the increases in product prices then trigger three sets of responses.

The first response is conservation by consumers, who cut their consumption as product prices increase in response to changes on the commodity market. The second is expectational response by consumers who, seeing an increase in spot market prices, respond by increasing their inventories in anticipation of future price increases. The third response, described so vividly by the editors of *The Economist*, is a rush by oil companies to build stocks.

Essentially, the increase on commodity markets, or the expectation of an increase, triggers an increase in demand for stocks, because consumers and refiners anticipate future increases in oil prices and wish to acquire supplies before this happens. This rush, which overwhelms the conservation response, is motivated by two institutional factors. First, suppliers of crude oil—both countries and companies—

historically have not responded instantaneously to changes in spot market prices. Second, marketers of petroleum products generally have based prices on the cost of crude, so that consumer prices increase slowly with crude costs, not quickly with spot prices. This creates an incentive to build inventories. Thus, the shift in supply starts a process that, for a time, feeds upon itself.

This price-setting process has been described by others, but it is quantified here. We find that producing countries follow a very regular pattern in responding to changes on commodity markets. Initially (i.e., during the first year), producers of crude oil extract only a small portion of the increase in value of crude, but eventually they capture the full amount of that increase in value.

The second institutional phenomenon concerns the way in which changes in crude prices are passed on in consumer prices. For a variety of reasons, changes in prices occur after OPEC increases crude prices, not after increases in spot market prices. This tends to minimize short-run conservation effects. Oil companies and consumers having the capacity to increase their inventories realize that an increase in commodity market prices represents a signal that crude oil prices and consumer product prices will soon rise. They therefore quite correctly perceive that it is in their best financial interest to "dash for excess stocks," as *The Economist* put it.

Once these two processes are understood, it becomes possible to analyze the proposed mechanisms to dampen the increase in commodity prices, deflate the sudden surges in demand for inventories, and thus lead to a more stable process of adjustment to increases in oil prices during a disruption. These mechanisms are suggested in later chapters.

## THE NATURE OF A DISRUPTION IN OIL SUPPLIES

A disruption in world oil supplies occurs when there is either a loss of reserves or a loss of producing capacity. The following examples suggest basic types of disruptions.

- The most severe type of disruption is a permanent physical loss of reserves in a major supply region, due either to an act of God (an earthquake, for instance), an act of war that leads to the destruction of producing facilities and causes an irreparable loss of potential oil recovery, or social disintegration, such as what occurred in Iran.

- A less severe disruption is a temporary or permanent reduction in production by one or more producing countries for any reason, including economic or political motives.
- Another type of disruption is a temporary or permanent disruption of the system by which oil is transported from producing to consuming regions.

The effect of any of these actions would be a leftward shift in the static supply curve[1] of oil, as shown in Figures 1-1 and 1-2.

Figure 1-1 illustrates the effect of a permanent physical loss in the world supply of oil. In this case, a portion of total oil reserves is lost permanently. This is precisely the sort of interruption that began in 1978 when a strike by oil workers in Iran started the process leading to the fall of the Shah. The political chaos caused the termination of a very elaborate enhanced oil recovery program designed to increase the volume of oil ultimately extracted from the Iranian oil fields. A significant portion of this increased production is now thought to be permanently lost. Thus, reserves in Iran have been lost.

The effect of a loss in reserves is demonstrated by the leftward movement of the vertical line labeled "Capacity Before Disruption." This represents the maximum sustainable productive capacity of the world's oil producers at any given time.

A loss in capacity may or may not affect the supply function for oil. If, for instance, the loss of capacity occurs in an area where oil is not now being produced, there may be no adverse effect. (This situation is shown in Figure 1-1a.) On the other hand, if the loss occurs

---

1. The discussion in this chapter assumes the existence of a supply curve for oil. To make this assumption, one must assume that the supply of oil on world markets is characterized by competitive conditions because, as every graduate student in economic theory learns, monopolists and imperfectly competitive industries do not have supply curves. It is not our intention to support such an assumption—indeed, we believe that even the most casual examination of the facts would lead one to conclude that the world oil market is not characterized by competition. Instead, we assume that at any moment of time the production and price of oil are controlled by a number of small suppliers and one very large supplier (Saudi Arabia) and that supplier acts in a fashion that might loosely be described as monopolistic and maximizing long-run profits. Thus, we assume that Saudi Arabia is willing at any moment in time to make modest adjustments in the volume of oil produced in order to maintain roughly stable prices.

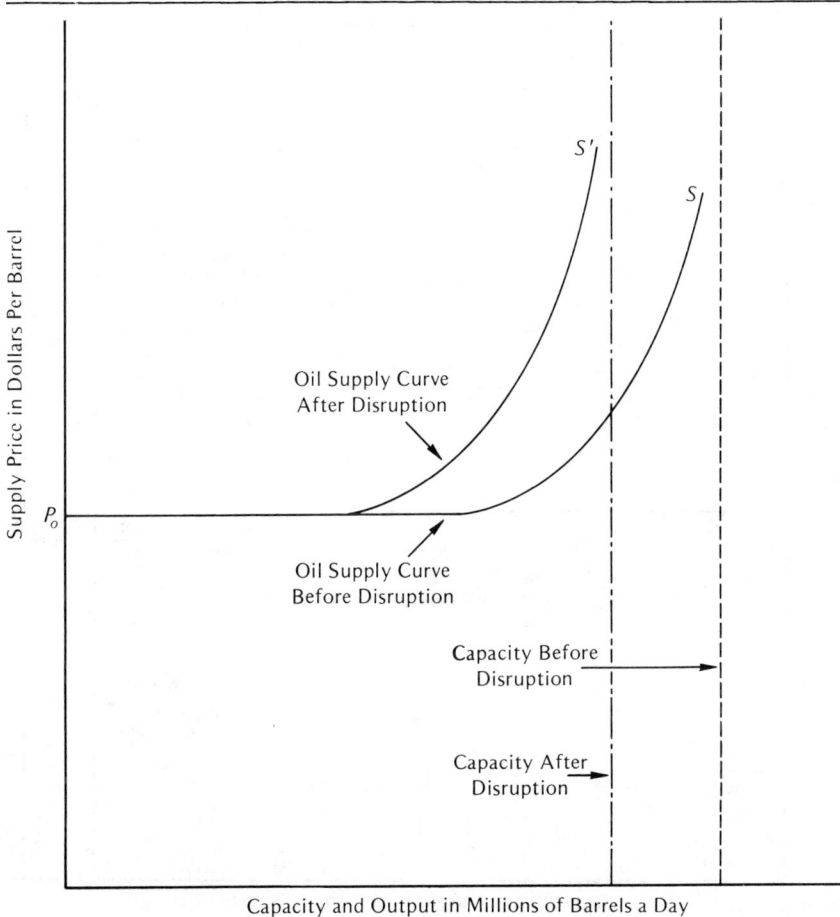

Figure 1-1. Static World Supply of Crude Oil.

in an area where production is currently taking place (as was the case in Iran), the loss will cause the leftward movement of the supply curve. This is the case shown in Figure 1-1.

A loss of reserves may influence some producers not directly affected by the loss, because a loss in reserves anywhere may alter every producer's optimal intertemporal producing plan (Pindyck 1978). This would occur if a loss of reserves in area $A$ altered expectations of long-term prices and changed the expected discounted value of future revenues in areas $B$ and $C$, causing these

**Figure 1-1a.** Static World Supply of Crude Oil.

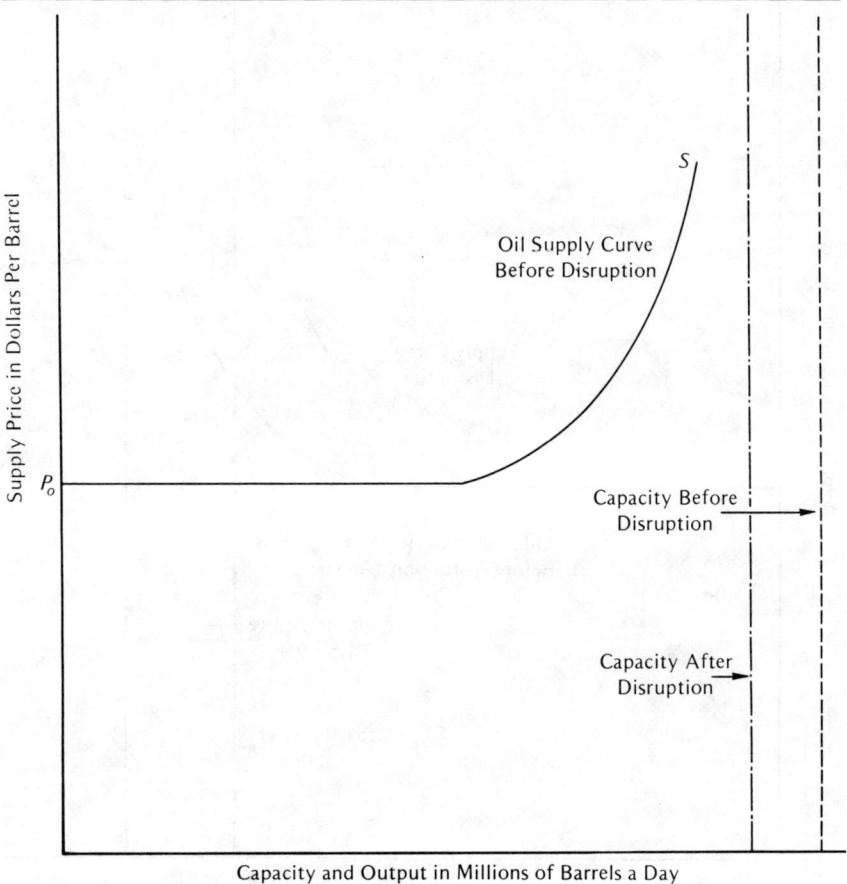

countries to defer production.[2] Thus, a leftward shift of the supply curve results if the loss of reserves occurs in a producing area or if it occurs in a nonproducing area but causes other producers to defer some part of their production in order to maximize future income.

2. Essentially, it is assumed that every producer establishes a production plan for current and future output (measured in terms of a sequence of output levels $Q_t$ ($t = 0$ to $\infty$)) based on an expectation of future prices, $P_t$, and on the producer's discount rate, $r_t$, so as to maximize the present discounted value of reserves

$$PDV = \sum_{t=0}^{\infty} \frac{P_t Q_t}{(1-r)^t}$$

**Figure 1-2.** Static Depiction of Reduction in Supply.

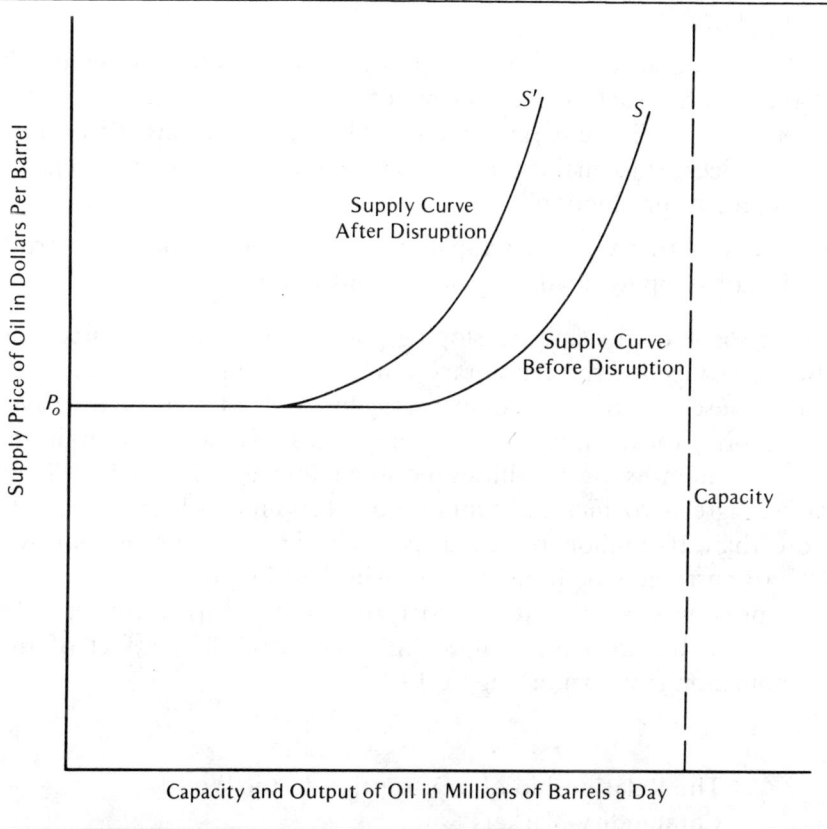

Capacity and Output of Oil in Millions of Barrels a Day

A different type of situation is illustrated in Figure 1-2. In this case, there is no loss in reserves or capacity. Instead, one or more producers elect to reduce the level of production. Such an action might be caused by:

- A political embargo directed against one or more customers like that imposed by the Organization of Arab Petroleum Producing Countries against the United States and the Netherlands in 1973,

---

A permanent loss of reserves in any country should lead to a change in the expectation of future prices (unless, that is, future prices are expected to be zero, i.e., the market will cease to exist). It follows that if expected prices in the future increase, a producer will want to trade current production for future production in order to maximize the present value of income under the new regime.

or the embargo directed against the United Kingdom by Nigeria in 1979: The embargo must be accompanied by a reduction in supply.

- A decision to reduce current production to preserve reserves for the future: Such a decision might occur if a producing country lowered its discount rate or changed its expectations as to future oil prices, thus making future production more valuable relative to present production.
- A loss of transportation capacity due to the temporary interruption of shipping, loading facilities, and so forth.

The shape of the upward sloping portion of the supply curve may also be changed if the temporary reduction in supply from one producer causes another to compensate, but only if the buyers pay a higher price for the incremental supply. This was the situation during the first months of hostilities between Iran and Iraq when Saudi Arabia agreed to increase supply from 8.5 million barrels a day to more than 10 million barrels a day. The increased production was sold to customers of Iran and Iraq who had lost their supplies. The sales price was set, not at the price the Saudis charged their regular customers, but at a price of $36.00 per barrel. The effect of this phenomenon is shown on Figure 1-3.

### The Relationship of a Disruption to the World Commodity Markets

A disruption in oil supplies can be caused by an action of either a producing country or a producing company. Although most recent attention has focused on actions by countries (Deese and Nye 1981), many of the studies written in the late 1960s and early 1970s focused on certain multinational companies that were accused of conspiring to hold back production.

Whatever the source of the loss of supply its effect is noted initially in the increase in prices of products traded on commodity markets for the following reasons. First, the buyers, typically oil companies whose supplies have been cut, may attempt to replace supplies by purchasing products on these markets before knowledge of the disruption becomes widespread. Second, these same companies may determine that inventories of products that previously had

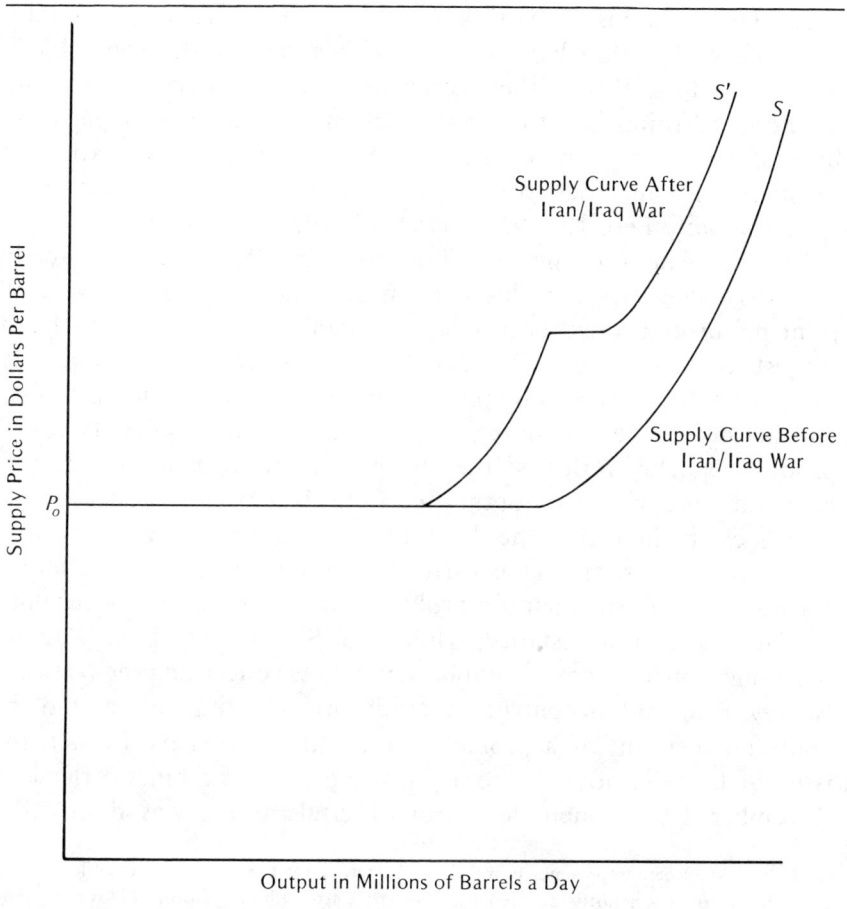

**Figure 1-3.** Change in Static Supply Curve for World Oil Illustrating Response to Saudi Sale of Incremental Production During Iranian Shutdown.

appeared to be surplus and therefore were slated to be sold on the spot market are no longer surplus. Third, those companies losing crude may cancel contracts with third-party buyers, forcing them to rush to the spot market. The net effect is to cause products that otherwise would have been sold on the commodity markets to remain in inventory. The result is a quick increase in the prices of products traded on the oil commodity markets: premium gasoline, regular gasoline, gas oil, jet fuel, naphtha, and fuel oil (residual fuel oil) of various sulfur grades.

The 1979–80 Iranian crisis offers a classic example of this process. The loss of Iranian supplies forced purchasing companies such as BP to cut back sales to third parties by approximately 700,000 barrels a day. One large loser was Exxon, which had a third-party contract to purchase 350,000 barrels a day of Nigerian crude from BP. To balance its loss, Exxon then curtailed its third-party sales and announced intentions not to renew third-party contracts with Japanese buyers. Thus, one consequence of BP's loss of crude was to force Japanese oil companies to turn to the spot market (*Petroleum Intelligence Weekly* Feb. 12, 1979; March 19, 1979).

The movement of product prices on the Rotterdam market in late 1978 demonstrates this phenomenon nicely. The prices of all principal products had been relatively stable through the months of August, September, and October. However, prices increased quickly in late October as strikes by pilots at the Kharg Island loading facility and in the southern export ports forced cuts in exports. Prices of products quickly reflected the changed situation, rising by 20 percent and more within a month. (See Table 1-1.)

It is clear, then, that the disruption was well under way and spot markets were responding as early as October 1978. The first public discussion of possible supply problems, however, did not occur until January 1979. For instance, while U.S. Secretary of Energy James Schlesinger made frequent public statements concerning the National Energy Plan and decontrol of crude prices during this period, he made no mention of a problem on world oil markets. In fact, the issue of Iran did not even come up at meetings of cabinet officials in December 1978, when decontrol of crude prices was discussed.[3]

---

3. This statement represents more than conjecture by the author. The subject of oil price decontrol was a hotly debated topic in the Carter administration in November and December 1978 because the law requiring that price controls be imposed on crude oil permitted the President to suspend them after June 1979. The discussion arose in late 1978 during the process of the preparation of budget proposals for fiscal year 1980. In connection with this process, meetings of the cabinet and White House officials involved in the discussion were held on December 1, 1978, December 15, 1978, and December 21, 1978. The Secretaries of Treasury and Energy were present at all meetings, as were the chairman of the Council of Economic Advisers, the President's anti-inflation adviser, and others. The author was present to take notes for the Secretary of the Treasury. According to these detailed notes, the subject of the developing Iranian crisis and its potential effect on oil prices never arose.

This view is reenforced by the administration's calculations of revenue projections from the windfall profits tax issued in late April 1979 when the decontrol decision was announced. Even as late as April, the White House envisioned year-end prices of only $16.00/bbl.

Table 1-1. Prices of Principal Petroleum Products on Rotterdam Market (*dollars per barrel*).

| Date | Gasoline (Regular) | Gasoil (#2 Fuel Oil) | Residual Fuel Oil (1% sulphur) |
|---|---|---|---|
| October 1, 1978 | 21.42 | 18.48 | 12.42 |
| November 1, 1978 | 23.94 | 21.00 | 15.70 |
| December 1, 1978 | 24.78 | 21.42 | 14.35 |
| December 15, 1978 | 23.94 | 20.58 | 14.04 |
| January 1, 1979 | 23.10 | 24.36 | 16.01 |
| January 15, 1979 | 25.20 | 26.46 | 17.52 |
| February 1, 1979 | 26.88 | 34.02 | 18.12 |
| February 15, 1979 | 35.70 | 42.84 | 21.29 |
| March 1, 1979 | 40.74 | 42.00 | 20.99 |
| March 15, 1979 | 32.34 | 34.02 | 18.57 |
| April 1, 1979 | 32.34 | 34.86 | 18.27 |
| April 15, 1979 | 35.70 | 37.38 | 19.33 |
| May 1, 1979 | 37.80 | 37.80 | 20.84 |
| May 15, 1979 | 40.74 | 39.06 | 21.29 |
| June 1, 1979 | 48.72 | 54.18 | 25.07 |
| June 15, 1979 | 45.36 | 51.24 | 25.67 |

Source: *Platt's Oil Price Service* (various issues 1978-79) and *Petroleum Intelligence Weekly* various issues 1978-79.

A similar pattern may be observed in the 1973 crisis, when Arab states embargoed the United States and the Netherlands and simultaneously acted to reduce supply. The immediate response was an increase in the price of products, with public discussion of the developing crisis occurring some days later.[4]

The explanation for a price increase on commodity markets during a disruption can be explained easily through reference to any of the figures offered in the previous section by the introduction of a demand curve. The reduction in supply causes the market clearing

---

4. It should be noted that a disruption probably also will induce increases in the spot market prices of crude oil. During an interruption the press frequently notes changes in spot market crude prices. These reports are, however, less reliable than the reports from the major commodity markets, because there is no formal market for crude oil and because crude oils vary widely in weight (gravity) and quality. Furthermore, terms of sale of crude will vary depending on market conditions and price, with two cargoes of like quality crude often selling the same day for different prices due to differences in payment terms.

**Figure 1-4.** Effect of Shift in Supply Is to Raise Price from $P_o$ to $P_1$ if Price Change Is Simultaneous with Shift in Supply.

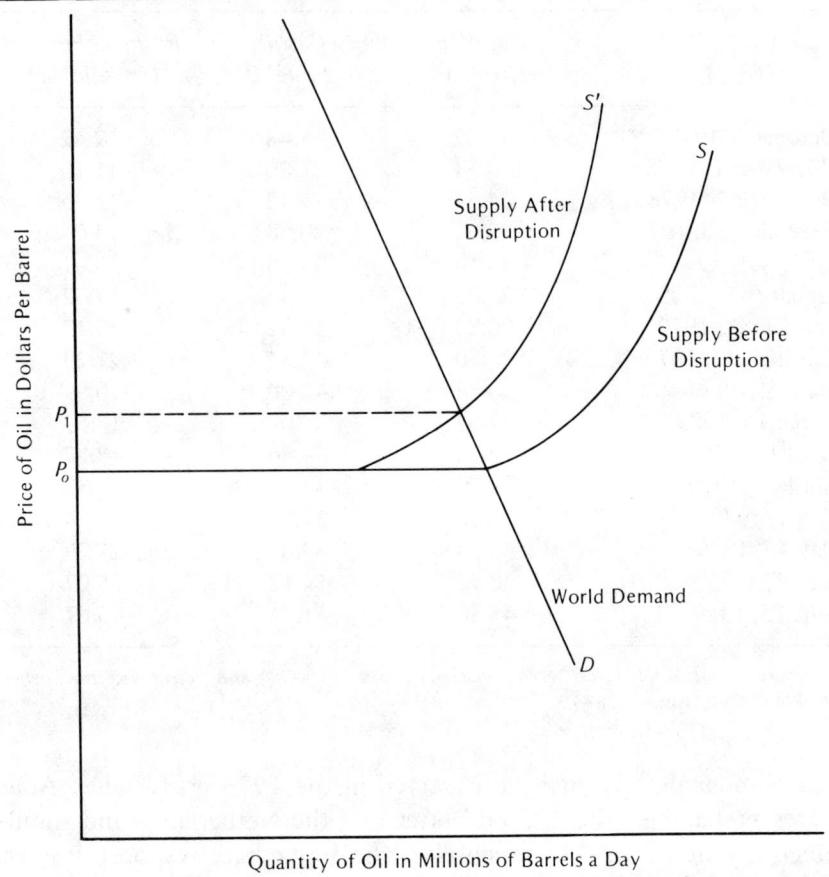

price to increase from $P_o$ to $P_1$ (see Figure 1-4). Thus the initial effect of an interruption on prices appears simple: The oil price rises from $P_o$ to $P_1$.

However, the process does not stop at this point because the price of crude oil does not initially increase to $P_1$. Instead, while there is an initial increase in the value of crude oil measured from commodity market prices of petroleum products from $P_o$ to $P_1$, the prices of products sold by refiners and marketers to consumers and the price of crude sold to refiners by producers remain at $P_o$. Therefore the initial effect of the shift in supply is to create a spread (or vacuum) which then triggers a new and different response.

## The Impact of an Interruption on the Demand and Price of Oil

> ... Economic collapse often has the character of a cumulative process. Let it go beyond a certain point and it will tend for a time to gain strength from its own development as its effects spread and return to intensify the process of collapse. (Friedman and Schwartz 1963.)

An interruption in world oil markets does not begin and end with the shift in the supply curve for two reasons. First, exporters do not immediately move prices from $P_o$ to $P_1$; they make initial adjustments of smaller magnitudes, creating transitory shortages because supply at prevailing prices is less than demand at those prices. Second, the world demand curve for oil does not remain fixed as it appears in Figure 1-4. It shifts, and the direction of the shift is perverse; that is, the demand curve shifts to the right during the first stages of a disruption as consumers increase their demand for stocks in anticipation of future price increases. The combination of these two phenomena raises the final equilibrium price above $P_1$ by a substantial margin.

The key determinant of this process is the manner in which crude oil prices are established by exporters, particularly the individual members of OPEC. The process can be described as one of slow adjustment and is shown in the data from the three disruptions experienced during the 1970s. In each case prices were increased in small increments in response to large changes on the Rotterdam market. This process can be observed by comparing spot product values (the value of a barrel of crude as determined by the prices of products derived from it) with official contract prices (see Table 1-2 and Figure 1-5).[5]

---

5. Spot values are also referred to as "netbacks." Spot values are computed for any type of crude oil by taking a weighted average of the prices of the products derived from a barrel of crude oil and subtracting refining and freight costs. Mathematically, the spot value, $SV$, can be represented as:

$$SV = \sum_{i=1}^{N} w_i P_i - (C + T),$$

where $w_i$ (the weights) represents the percentages of principal petroleum products derived from a crude oil in distillation, $P_i$ the prices of these products, $C$ the marginal cost of refining, and $T$ the cost of transportation. The calculation of spot values for a particular type of crude oil are shown on Table 1-2. The weights are typically described as coming from a hydro-skimming refinery. See the Appendix to this volume for details.

**Table 1-2.** Spot Product Values (Netbacks) and Official Prices of African Light Crude, 1978 to 1980 (*dollars per barrel*).

| Date | Spot Product Value | Official Contract Price |
|---|---|---|
| **1978** | | |
| January | $13.71 | $14.20 |
| February | 13.50 | 14.20 |
| March | 13.88 | 14.20 |
| April | 14.13 | 13.95 |
| May | 14.29 | 13.95 |
| June | 14.06 | 13.95 |
| July | 14.14 | 13.90 |
| August | 14.82 | 13.90 |
| September | 14.82 | 13.90 |
| October | 15.42 | 13.95 |
| November | 18.23 | 13.95 |
| December | 16.79 | 13.95 |
| **1979** | | |
| January | 19.71 | 14.73 |
| February | 27.08 | 14.87 |
| March | 25.49 | 15.24 |
| April | 27.65 | 18.42 |
| May | 33.52 | 19.54 |
| June | 37.67 | 21.07 |
| July | 34.35 | 23.41 |
| August | 33.18 | 23.41 |
| September | 34.09 | 23.41 |
| October | 35.79 | 24.19 |
| November | 39.45 | 26.18 |
| December | 40.49 | 28.06 |
| **1980** | | |
| January | 38.32 | 33.10 |
| February | 35.80 | 35.45 |
| March | 34.82 | 35.45 |
| April | 32.27 | 35.65 |
| May | 35.98 | 37.25 |
| June | 34.83 | 37.25 |
| July | 34.10 | 38.15 |
| August | 31.58 | 38.15 |
| September | 31.58 | 36.90 |
| October | 34.25 | 36.90 |
| November | 37.18 | 36.90 |
| December | 34.97 | 36.90 |

Source: *Petroleum Intelligence Weekly* (Special Supplement Feb. 2, 1981).

ECONOMIC CHARACTERISTICS OF AN INTERRUPTION

**Figure 1-5.** Contract (or Posted) Price of Crude (Broken Line) Follows the Spot Value (Solid Line).

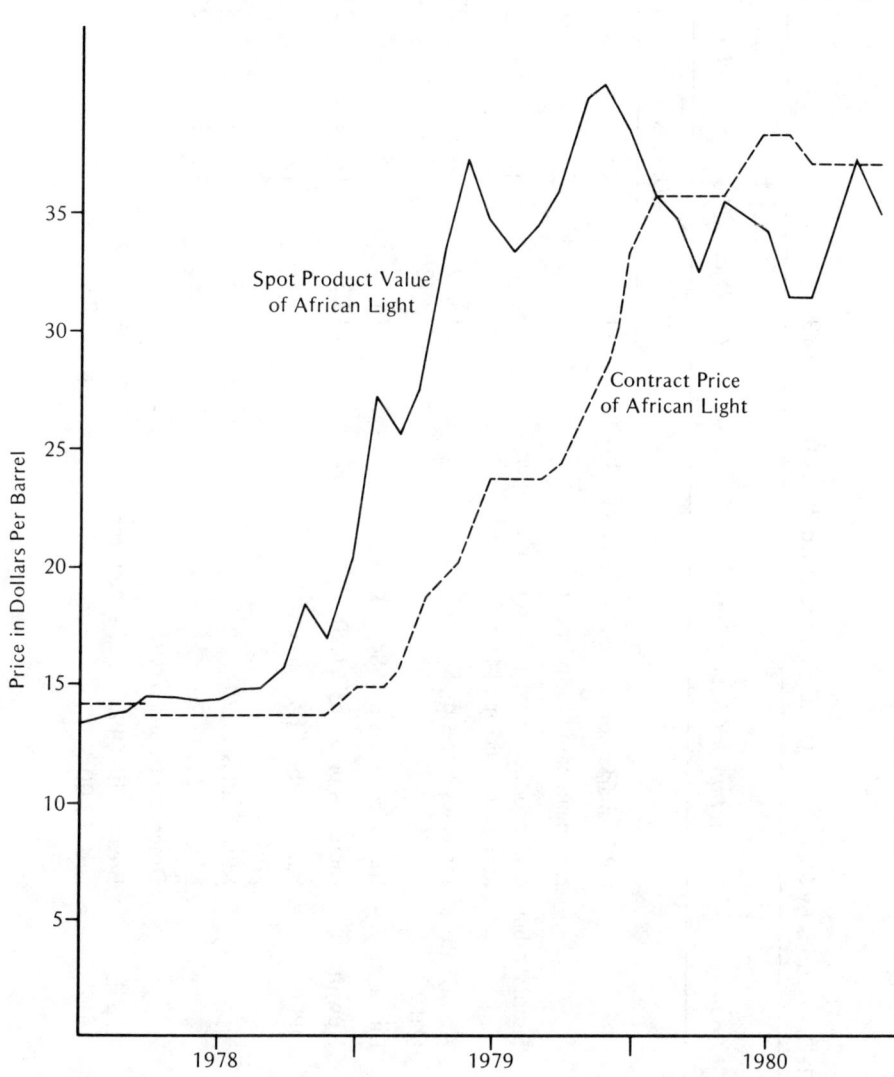

Source: *Petroleum Intelligence Weekly.*

Table 1-3. Actions by Members of OPEC that Increased Prices During 1979.

| Date | Country | Action or Circumstance | Old Price | New Price | Estimated Spot Value at Time of Increase |
|---|---|---|---|---|---|
| | | | Dollars per Barrel | | |
| January 1 | | OPEC marker price established at $13.33 | $12.70 | $13.33 | |
| January 15 | | European panic for product | | | $17.00 AM |
| February 5 | Saudi Arabia | Cuts production; sells incremental crude at higher price of $14.55 | 12.70 | 14.55 | 22.21 AL |
| February 19 | Abu Dhabi | Increases prices by 84¢ to $1.02 | 13.77 | 14.71 | 22.21 AM |
| | Qatar | "   "   "   " | 13.78 | 14.62 | 22.21 AM |
| February 26 | Libya | Increases prices by 68¢ | 14.52 | 15.20 | 27.08 AL |
| March 5 | Kuwait | Increases prices by $1.20 | 13.39 | 14.52 | 22.22 AH |
| April 2 | OPEC | Saudi Arabia lifts marker for all production | 14.55 | 14.54 | 22.13 AM |
| | | Kuwait adds $1.20 | 14.52 | 15.73 | 22.22 AH |
| | | Nigeria adds $3.70 | 14.82 | 18.52 | 25.29 AL |
| | | Others up by similar amounts | | | |
| May 28 | Algeria | Breaks with OPEC price structure, adds $3.00 | 18.46 | 21.00 | 33.52 AL |

| Date | Country | Action | | | |
|---|---|---|---|---|---|
| July 2 | OPEC | Second 1979 meeting | 14.56 | 18.00 | 31.76 AL |
| | | Kuwait adds $3.06 | 16.93 | 19.99 | 28.56 AH |
| | | Nigeria adds $2.49 | 21.00 | 23.49 | 37.67 AL |
| | | Algeria adds $2.50 | 21.00 | 23.50 | 37.67 AL |
| August 13 | Kuwait | Increases prices by $2.00 on small volumes | 19.99 | 21.99 | 27.97 AH |
| October 15 | Kuwait | Applies increase to all crude | | | |
| October 22 | Iran | Raises prices by $1.50 retroactive to 10/79 | 22.00 | 23.50 | 30.23 AM |
| | Libya | Raises prices by $2.77 | 23.28 | 26.05 | 35.79 AL |
| October 29 | Algeria | Follows Lybia | 23.50 | 26.27 | 35.79 AL |
| November 13 | Nigeria | Follows Lybia | 23.49 | 26.26 | 39.45 AL |
| December 17 | Saudi Arabia | Adds $6.00 prior to meeting of OPEC ministers | 18.00 | 24.00 | 34.65 AM |
| December 31 | Libya | Adds $3.73 | 26.05 | 29.78 | 40.49 AL |

Source: *Petroleum Intelligence Weekly* (issues indicated by date in first column).

Note: AM = Arab marker.
AL = African light.
AH = Arab heavy.
SV = Spot value.

The process by which the prices of crude oil sold under contract follow spot values is examined statistically in the next chapter where we demonstrate that there is a close dynamic link between official prices and spot values. However, the linkage can be demonstrated more dramatically by analyzing the chronology of price increases imposed during 1979 by the various OPEC states. This is roughly shown on Table 1-3, where movements of official prices charged by many countries are compared with reported spot values for those countries.

The cycle of price increases in 1979 began in January when members of the cartel introduced the first of three contemplated quarterly price increases that had been announced at the previous meeting in December. Saudi Arabia increased the official per barrel price of Arab light from $12.70 to $13.33 a barrel. However, at the same time the spot value of Arab light (the netback) had risen to $17.00 on European commodity markets as consumers scrambled for product. Then, at the end of January Saudi Arabia announced a reduction in production for the months of February and March and a price increase on small volumes of crude. The effect of this announcement was to drive spot values to $22.21.

In response to high product prices on European commodity markets, a few of the more venturesome members of the cartel began to raise their postings. Abu Dhabi and Qatar increased their prices by $0.84 to $1.02 during the first weeks of February.

Early in March Kuwait responded to continuing high spot market values by raising its prices by $1.20 from $13.39 to $14.52. Then, at the end of March the members of the cartel met to review the pricing picture. At that time the spot values of crudes ranged from $22.13 for Arab light to $25.29 for African light while official prices charged by various exporters were in the $13.00 to $14.00 range. This large price spread caused OPEC to announce another price increase. The increase did not eliminate the spread, however, because prices were increased by only between $1.20 and $3.70.

In the ensuing months spot values continued to increase and periodically official prices changed. Sometimes the price changes were announced unilaterally, while twice the changes resulted from meetings of the cartel. In no case, however, did the OPEC member countries attempt to bring official prices up to the level of spot values.

Thus, this analysis indicates that the members of OPEC have followed the commodity markets in setting prices. This proposition is further demonstrated in Chapter 3.

## THE RELATIONSHIP BETWEEN PRICE DETERMINATION, DEMAND, AND EQUILIBRIUM PRICES

The preceding sections have described a disruption as characterized by a movement in the supply curve, shown that the disruption is first observed on the world commodity markets where spot values rise from $P_o$ to $P_1$, and shown that changes in prices on commodity spot markets eventually lead to increases in official crude oil prices. It is the lag between spot values and official crude prices that drives prices even higher. What appears to happen is that both consumers and refiners increase rather than decrease their demand for petroleum—either products or crude—at the first sign of a shortage. This can be shown in a simple supply diagram as a temporary rightward shift of the demand curve at the start of the disruption.

This type of behavior is shown in Figure 1-6. At the start of the disruption supply shifts to the left from the supply curve $S$ (labeled "Supply Before Disruption") to the supply curve $S'$ (labeled "Supply After Disruption"). This shift induces the spot value of crude oil to increase from $P_o$ to $P_1$. However, the prices paid by most purchasers of crude oil, such as the multinational oil companies, remain at approximately $P_o$, creating a temporary shortage in the crude market, the size of which could be equal to the difference between $Q_o$ and $Q_1^*$.

This shortage appears to create a dilemma for purchasers of crude oil. They could, if they desired, raise product prices to reflect the increase from $P_o$ to $P_1$, thereby eliminating the shortage in the markets for petroleum products. This action would enable the buyers of crude to temporarily capture the profit created by the slow adjustment in crude oil prices.

Alternatively, the purchasers of crude oil could refrain from increasing prices—or be restrained by public policy in consuming countries. In that case, prices would remain at $P_o$, and there would be a shortage equal to at least $Q_o - Q_1^*$. However, it is arguable that the gap may be much larger than the difference between $Q_o$ and $Q_1^*$. If the supply of crude oil is reduced to $Q_1^*$ but prices are initially held to $P_o$, buyers may elect to distribute only part of the supply—the volume labeled $Q_1$ in Figure 1-6. If this type of behavior occurs, the spot value will then be driven to $P_2$, not $P_1$, and the temporary shortage will be represented by the difference between $Q_o$ and $Q_1$.

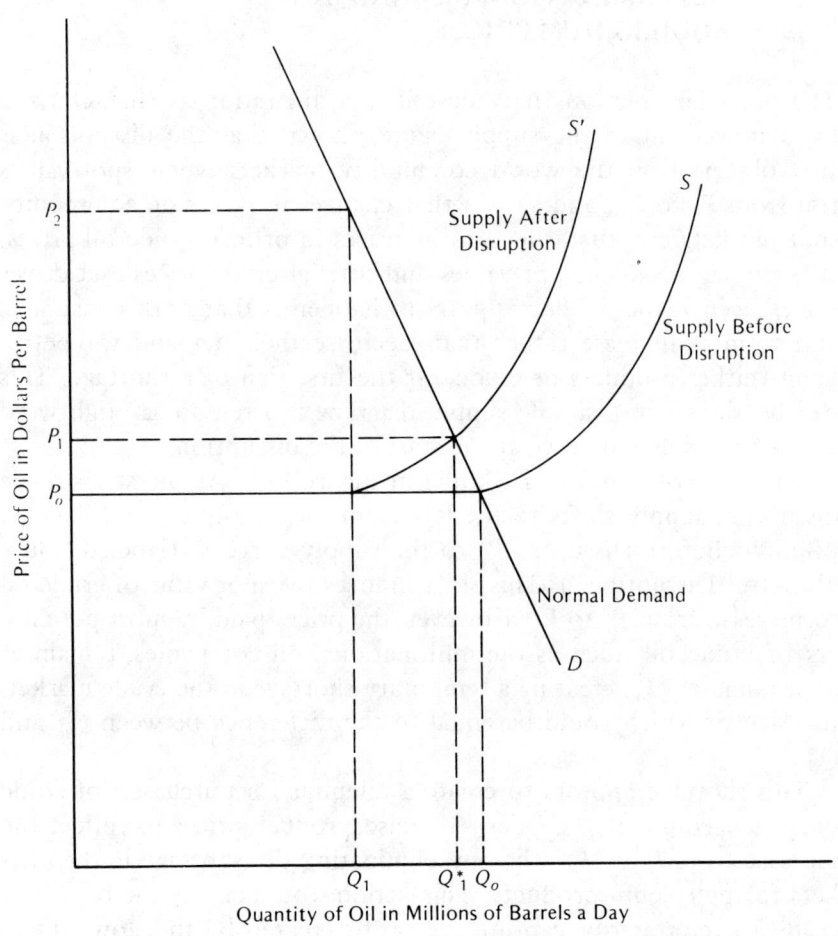

**Figure 1-6.** Effect on the Short-Run Equilibrium Price of a Failure by OPEC to Raise Prices Is to Drive that Price Higher.

Behavior of this sort could occur if purchasers of crude oil understood the process by which official crude prices are established and were either unable or unwilling to immediately raise their product prices but were able to increase their inventories of crude oil or products. In fact, this appears to be exactly what happens in a crisis.

Consumers also may increase their stocks of petroleum products in response to differences between spot values and the price of crude. Indeed, consumers with the capacity to stockpile oil inventories would be foolish not to attempt to increase their stocks as soon as spot values begin to increase, especially if they can acquire the additional volume of products at producer prices rather than the much higher spot prices.

The effect of consumer stockpiling at the start of a disruption may be shown graphically by shifting the demand curve to the right in a simple supply/demand diagram. This is shown in Figure 1-7. Such a response has the effect of driving equilibrium prices even higher than before, from $P_2$ to $P_3$. (Of course, during a glut an opposite response might be expected, with consumers reducing stocks in anticipation of future sales.)

Unfortunately, the available statistics on consumption do not separately identify consumer stock building and consumption. It is noteworthy, however, that observers of the oil market found in the spring of 1979 that consumption in the United States in the months of January, February, and March had been much greater than might have been expected, given the economic and climatic conditions at the time.

Most observers have been at a loss to explain the surge. However, if it is noted that consumption really measures only "disappearance from primary supply,"[6] the surge in demand can be explained as a response by consumers who, seeing the high prices on the spot market, attempted to acquire greater stocks. Such behavior is very similar to that observed in consumer markets where individual consumers rush out to acquire goods such as liquor or cigarettes on the eve of a tax increase. It is also observed in other markets, such as steel, where companies will stockpile inputs to production in advance of a price increase.

It may also be noted that consumers will reduce stocks during periods of apparent surplus in the world oil market. Such behavior would help to explain the apparent increase in conservation that occurred in the United States and other areas during the early

---

6. Oil statistics for the United States and most of the members of the Organization for Economic Cooperation and Development (OECD) report on changes of inventories held by refiners and terminal operators and on sales to ultimate consumers. They do not, however, provide any indication as to changes in stocks held by consumers, even though these stocks may be large in the case of some consumers such as manufacturers or electric utilities. Thus, the sale of 100,000 barrels of oil to a utility will be reported as consumption, even though the oil may only move across the street from a refiner's tank to a utility's tank. Should the utility decide to increase inventories, reported consumption will increase; should it decide to reduce stock holding, reported consumption will decline.

**Figure 1-7.** Effect on Short-Run Equilibrium Price of Consumer Hoarding Response Is to Drive Prices Even Higher.

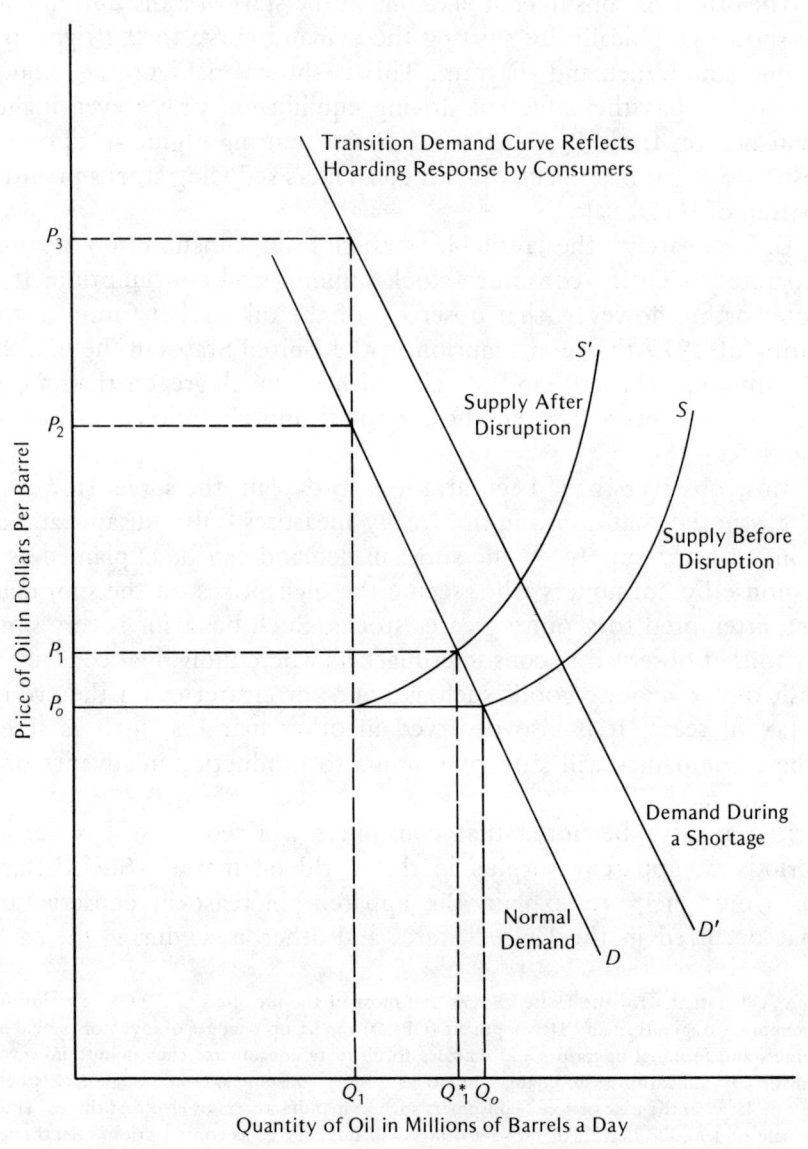

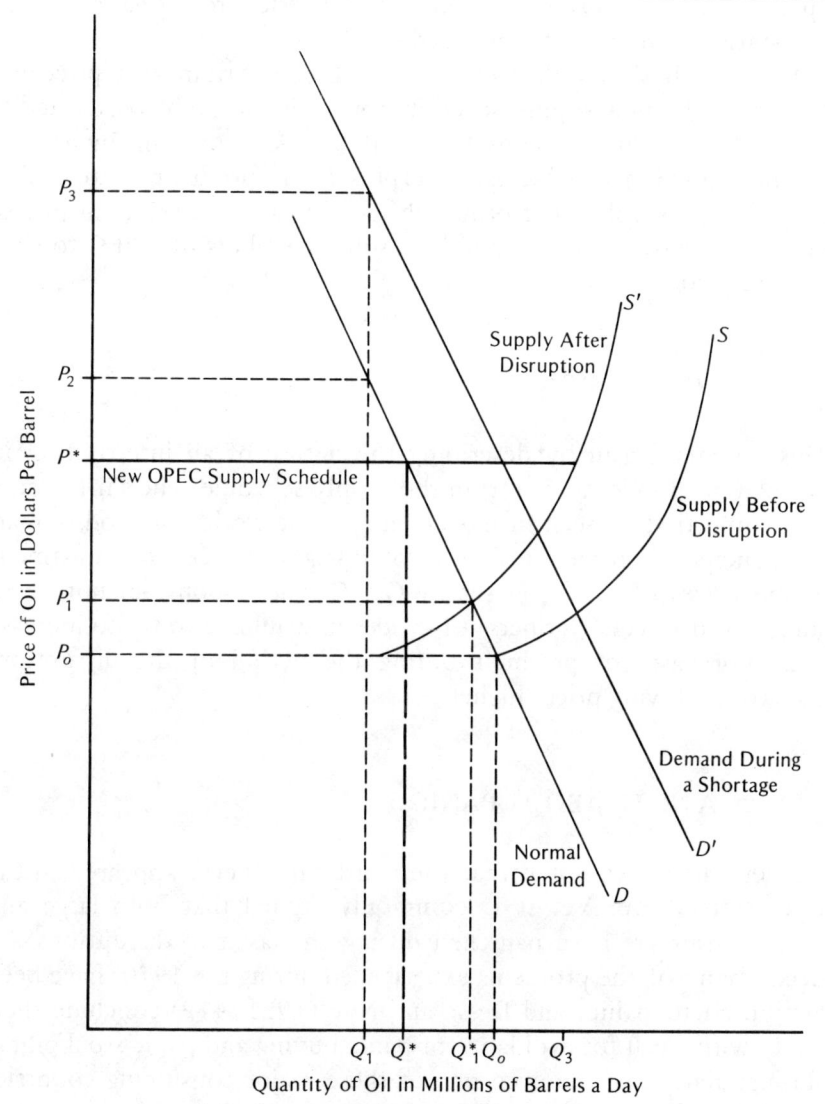

**Figure 1-8.** Equilibrium Is Restored when OPEC Raises Floor Price from $P_o$ to $P^*$.

months of 1981 when apparent consumption (disappearance from final supply) fell well below the levels that might be predicted, given prices and the level of income. Thus, during periods of glut we expect that consumers will reduce their purchases of oil and tend to use up their stocks, causing the data on deliveries from final stocks to overstate the amount of conservation.

Figure 1-8 shows the restoration of equilibrium at a price $P^*$. Note that the new supply schedule for oil is infinitely elastic at $P^*$, so long as demand remains less than $Q_3$. As shown in Figure 1-8 the new price level $P^*$ creates surplus crude-producing capacity in the world, a surplus that ordinarily would cause a decline in prices, but, in the case of oil, may induce some members of OPEC to shut in production.

## CONCLUSION

This chapter began by describing the nature of an interruption in petroleum supplies as a shift in the supply schedule. The shift in supply would first be seen in movements on the world commodity markets. Increases on these markets would begin a process of adjustment in the prices of crude oil set by OPEC. The response of both consumers and buyers (refiners) of crude oil would be to try to increase their purchases of products during the period of the disruption, effectively driving prices higher.

## A NOTE ABOUT PANIC

Readers may have noted that the word panic never appeared in the discussion above. Yet it is commonly argued that both large and small consumers have panicked during the last two disruptions. Indeed, many of the problems experienced during the 1970s have been attributed to panic, and Desse and Nye (1981: 419) conclude their study with a call for avoidance of panic buying and panic stockpiling, stating that "panic in even one of these major consuming countries could disrupt the entire market system."

In the discussion offered here, however, panic is replaced by a measurable phenomenon—the difference between the cost of crude and spot market values. When this difference increases, we expect consumers and oil companies alike to demand greater stocks. When the difference becomes very large, we anticipate that demand for stocks will become very large. Does such behavior amount to panic, or just rational behavior in light of expectations? We offer no answer. Instead, it is suggested in later chapters that policies purporting to solve the problems created by a disruption in world oil supplies must squarely face *and alter* the economic behavior that occurs during a crisis if policymakers desire to soften, rather than compound, the crisis.

# 2 THE BEHAVIOR OF OIL PRICES ON COMMODITY MARKETS DURING CRISIS

*Lucy*: You know why that big black bug doesn't move? Because she's the queen bug. She just sits there, see, while the other bugs do all the work.

*Charlie Brown*: That's not a bug—that's a jelly bean.

*Lucy*: By golly, you're right, Charlie Brown. I wonder how a jelly bean ever got to be queen?

Charles M. Schultz, as quoted in F. M. Fisher,
*A Priori Information and Time Series Analysis*
(1966: 3).

The economic theory of supply and demand is predicated upon the assumption that the incremental source of supply sets the market clearing price of a product. This theory began with Adam Smith and has been found in almost every economics text published, from Marshall to Samuelson. According to this theory, the incremental supply—or the market where incremental units are traded—represents the "queen bug" and sets the price or determines (causes) the market clearing prices. In the oil market, the Rotterdam market, together with the commodity markets in Singapore and the U.S. Gulf Coast, represents the incremental source of supply and thus should set the price of oil. Indeed, as shown in Chapter 3, Rotterdam does set the price of oil.

This view is not universally held, however, as demonstrated by the following quote.

> The nine nation European Community has started revealing key new information on members' inland consumer oil prices. Previously confidential, the figures have been collected for several years in an aim for "price transparency." ... The unstated goal is to "lay the ghost of Rotterdam to rest, permanently," European government officials tell *P.I.W.* They hope OPEC will get the message. (*Petroleum Intelligence Weekly* Aug. 6, 1979.)

The importance of the commodity markets, and by implication the traders who use them, was stated clearly by Roeber (1979: 50):

> The core function of traders is a balancing one. No country, company or refinery is ever in balance between supply and demand for all products, and an interchange to balance up surpluses and deficits is constantly taking place. By far the greatest part of this balancing takes place within and between the integrated systems, but the independent traders (at Rotterdam) play a part: (i) to take product from or supply it to the integrated systems at the margin, when main volumes have been balanced, ... (ii) to take speculative positions, term and spot, to buy for stock, speculating on seasonal price movements, (iii) to mobilize supplies from outside the major integrated supply systems, (iv) to act as a source for non-integrated downstream outlets.

To describe the behavior of petroleum markets during disruptions, one must go beyond the prices on the Rotterdam market and ask the question, "What causes movement of prices on spot markets?" Thus, the question addressed in this chapter is, "What moves Rotterdam?"

## THE BEHAVIOR OF PRICES ON COMMODITY MARKETS DURING DISRUPTIONS

The movement of prices on commodity markets during periods of uncertain supply is almost always in one direction—up. However, the rate and duration of the increase has differed in each of the three oil market disruptions experienced since 1973. During the 1973 embargo, prices rose dramatically, reached a peak 200 percent above beginning levels within two months after the outbreak of the war, and then quickly dropped off again by mid-February as uncertainties as to the volume of supply were settled. In the 1978-79 crisis, the initial response of prices was modest until exports were completely cut off on December 26, 1979. The sharp increase in prices did not occur until late January 1979, a full three months after the crisis began. However, because nagging uncertainties hung over the oil market for months, the increase in spot values in the end was greater after the Iranian crisis than after the Arab embargo. In the third episode—the war between Iran and Iraq—the initial surge in prices was identical to the 1973 episode and worse than the 1978-79 disruption, but within a week it became clear that war would have no lasting effects on supply.

The three experiences can be compared by examining the week-to-week increases in spot values. Figure 2-1 provides an illustration of this pattern.

Four factors appear to affect the manner in which prices change during a disruption: (1) the magnitude of the supply loss; (2) the speed with which supply is lost; (3) consumer or trader expectations as to the conditions of inventories at the time of the supply loss; and (4) the availability of alternative sources of supply. Based on an examination of each crisis, it is apparent that price increases are more likely to be large and precipitous when the loss of supply is abrupt than when it is gradual, when inventories are low rather than full, and when it is not possible to develop alternative sources of supply quickly.

This may be demonstrated by an examination of prior supply interruptions. These are the 1973 Arab Embargo, the 1980/81 war between Iran and Iraq, and the 1979 Iranian revolution. For convenience the Iranian episode is treated as four events: a gradual loss of supply experienced between November and January; a sudden loss of supply beginning at the end of January and lasting until April; a minor disruption beginning in May when a new, more conservative, Saudi policy was announced;[1] and a second sudden loss in supply occurring in November 1979 when the American embassy was seized.

To facilitate comparison of the six episodes identified here, the important information about supply and prices before, during, and after the disruptions is tabulated on Table 2-1. The information given there includes the month the disruption began, spot values of Arab light crude oil just prior to the start of the disruption; spot val-

---

1. The May episode also was triggered by a demand side measure taken when the United States offered a bounty (offered through the entitlement system) for imports of number two heating oil. This action points out that the spot market can be disrupted either by a supply side action or by a demand side action.

**Figure 2-1.** Market Response to Supply Disruption in Three Past Crises.

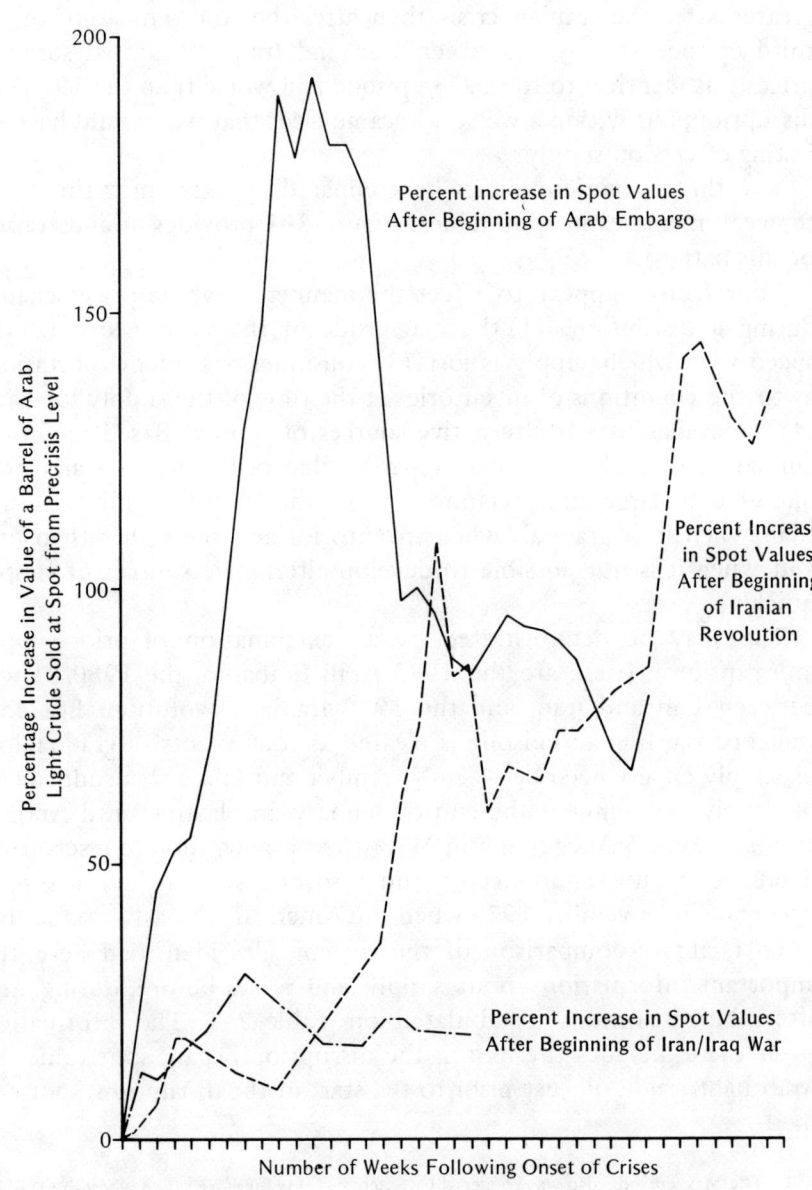

Source: Calculations by the author from data published in *Platt's Oil Price Service* (various issues).

ues thirty days after the start of the disruption; spot values sixty days after the start of the disruption; free world crude production at the beginning of the disruption; average free-world production during the first, second, third, and fourth months of the disruption; and the number of days of supply of stocks in the Organization for Economic Cooperation and Development (OECD) countries at the start of the disruption. (Days of supply are computed as the ratio of inventories in OECD countries to consumption during the month prior to the start of the disruption.)

Table 2-1 presents almost all the information available on each of the episodes of oil market disruption in a compact form that is rather difficult to interpret. Therefore, these data are summarized on Table 2-2.

Episode 1 began in mid-October 1973 when, in response to the outbreak of war between Israel, Egypt, and Syria, the members of the Organization of Arab Petroleum Exporting Countries (OAPEC) announced their intention to embargo shipments of oil to countries supporting Israel—primarily the United States. To implement the embargo, they also imposed cutbacks in oil production. As a result, free world production declined from 47.8 million barrels a day during September to 46.2 million barrels a day in October, and then to 43.2 million barrels a day in November. These cutbacks were gradually rescinded in December and January. Spot values of Arab Light on the Rotterdam market increased from $6.71 a barrel on the first of October to $11.53 per barrel on the first of November and $19.35 on the first of December as a consequence of the cut in production before beginning to give ground in January (when they fell to $18.19 a barrel) and February (when they dropped to $13.06).

There are two obvious reasons why the embargo had such a large impact on spot values. First, the production cutback was sudden and dramatic. The sudden nature probably contributed to substantial uncertainty in the market. Second, oil stocks were low. Inventories at the start of the crisis in those OECD countries for which data are available amounted to only sixty-five days of supply in September. This is an extremely low value for the early fall, especially in light of the seasonal surge in consumption which is expected to occur during the winter.

Episode 2 began in late October or early November 1978 when Iranian oil workers began a series of sporadic strikes, which eventually created the national turmoil in Iran that culminated in the fall of

Table 2-1. Comparison of Salient Characteristics of Six Episodes.

| Crisis | Began | Spot Value | | | Percentage Increase | |
|---|---|---|---|---|---|---|
| | | Start | 30 Days Later | 60 Days Later | 30 Days Later | 60 Days Later |
| | | Dollars/Barrel | | | | |
| 1973 Embargo | October 1973 | $ 6.71 | $11.53 | $19.35[a] | 71.8% | 188.4% |
| Iran I | October 1978 | 15.36 | 17.33 | 16.84 | 12.8 | 9.6 |
| Iran II | Late Jan. 1979 | 19.09 | 31.40 | 25.04 | 64.5 | 31.2 |
| Iran III | May 15, 1979 | 28.09 | 34.47 | 34.72 | 22.7 | 23.6 |
| Iran IV | November 1979 | 33.27 | 37.09 | 37.61 | 11.5 | 13.0 |
| Iran/Iraq | October 1980 | 30.38 | 38.23 | 35.84 | 25.8 | 18.0 |

a. Estimate.

Sources: Central Intelegence Agency, PIW, D.O.E., Platt's Oil Price Service, Organization for Economic Cooperation and Development.

the Shah. The initial strikes were small, but they were accompanied by an unrelated loss of 500,000 barrels a day of productive capacity in Iraq due to a pipeline fire. As a result of these two seemingly small events, free world production dropped by 4 million barrels a day (from 62.9 to 58.9 million barrels a day) over a period of two months. However, since the pattern of reduction in supply was erratic, spot values did not increase by large amounts. The value of Arab light rose from $15.36 to $17.33 per barrel between October and November, and then fell back to $16.84 during December. One reason for this slow increase in prices may have been stock drawdowns. Stocks in the month prior to the start of the strikes amounted to eighty-two days of supply and fell quickly to sixty-nine days of supply by January.

Episode 3 started on January 20, 1979, when Saudi Arabia announced a sudden cut in production. From November through January Saudi oil fields had been operating at maximum feasible rates as the companies operating there attempted to offset the loss in Iranian production. This action forced Saudi production to 9.8 million barrels a day, well above the ceiling of 8.5 million barrels a day established by the government. Then, in late January, Saudi officials announced that the 8.5 million barrel a day ceiling would apply for the entire first quarter of 1979. This meant that output would have

Table 2-1. continued

| Month Prior | At Start | Production Months After Start | | | | Duration | Stocks as Percentage of Consumption at Start |
|---|---|---|---|---|---|---|---|
| | | One | Two | Three | Four | | |
| | | Million Barrels a Day | | | | | |
| 47.8 | 46.2 | 43.2 | 43.6 | 44.2 | 45.4 | 4 mo. | 65 |
| 62.8 | 62.9 | 62.1 | 58.9 | 60.4 | 61.9 | 2 mo. | 82 |
| 60.4 | 58.9 | 60.4 | 61.9 | 62.5 | 62.4 | 2 mo. | 69 |
| 62.5 | 62.4 | 62.4 | 63.7 | 63.3 | 62.7 | 1 mo. | 72 |
| 63.3 | 63.1 | 62.6 | 61.7 | 62.2 | 61.7 | 2 mo. | 85 |
| 59.5 | 58.6 | 55.9 | 56.6 | 58.0 | 60.0 | 2 mo. | 120 |

to be reduced below 8 million barrels a day for February and March 1979. Such an action, if taken, would have had the effect of suddenly removing almost 1 million barrels a day from the market—an action similar to the production cut which accompanied the 1973 embargo. (*Petroleum Intelligence Weekly* Feb. 5, 1979).

Saudi production was cut for three or four weeks and world production dropped by approximately 500,000 barrels a day during February. The production cut was then restored in March. The effect on prices was dramatic. The spot value of Arab light crude oil increased 64.5 percent from $19.09 to $31.40 a barrel during the month of February. Then, as the crisis passed, the values dropped back to $25.04 a barrel by the end of March.

The effect of the Saudi temporary cutback appears to have been large in spite of the relatively small change in production for two reasons. First, the cutback was sudden. Second, it occurred at a time when stocks were low and worldwide production did not meet demand. Indeed, as *PIW* noted (Feb. 5, 1979: 5), "A cutback in the Saudi rate inevitably will force a faster drawdown in already rapidly dwindling world oil stocks. It is the difference between the current tight supply and an acute shortage."

Episode 4 occurred in mid-May 1979. It was a combination of a demand-side event and a supply-side event. The demand-side event was caused by an action of the U.S. Department of Energy, which effectively granted a bounty of $5.00 a barrel to importers of heating oil. This action sent ripples through the spot market and drove spot

Table 2-2. Comparison of Percentage of Change in Price and Output During Six Episodes.

| Episode | Cumulative Percentage Change in Spot Value | Cumulative Percentage Change in Output | Days of Supply at Start |
|---|---|---|---|
| Embargo | | | |
| Month 1 | 71.8% | -3.3% | 65 |
| Month 2 | 188.4 | -9.8 | |
| Month 3 | 171.1 | -8.9 | |
| Month 4 | 94.6 | -7.5 | |
| Iran I | | | |
| Month 1 | 12.8 | +0.2 | 82 |
| Month 2 | 9.6 | -1.1 | |
| Month 3 | 24.3 | -6.3 | |
| Iran II | | | |
| Month 1 | 64.5 | -2.5 | 69 |
| Month 2 | 31.2 | 0.0 | |
| Month 3 | 37.2 | +2.5 | |
| Iran III | | | |
| Month 1 | 22.7 | -0.2 | 72 |
| Month 2 | 23.6 | -0.2 | |
| Iran IV | | | |
| Month 1 | 11.5 | -0.3 | 85 |
| Month 2 | 13.0 | -1.1 | |
| Iran/Iraq | | | |
| Month 1 | 25.8 | -1.5 | 120 |
| Month 2 | 18.0 | -6.1 | |

Source: Table 2-1.

values up by $1.00 to $2.00 within twenty-four hours. The supply-side event was the uncertainty created by Saudi Arabia's announcement of its intention to sell more oil by direct sales and less through Aramco partners. *PIW* noted (May 14, 1979: 2), "Current liftings of the four U.S. companies have been trimmed about 300,000 to 400,000 barrels daily to some 6.8 to 7 million barrels a day from 7.2 to 7.3 million recently. There has been no official notification to Aramco of a cut in its entitlements, and reports of a reduction to

6.5 million b/d are premature, though directionally correct, *PIW* understands."

The combination of the U.S. bounty and the Saudi change was sufficient to tighten the market noticeably, although the data on world supply show little change. The spot value of Arab light increased from $28.09 a barrel in early May to $34.47 in early June and to $34.72 in early July.

Episode 5 occurred in November 1979 when the U.S. embassy in Tehran was seized by student militants. This seizure led to an embargo on imports of Iranian oil into the United States, an event that by itself would have had little effect on the market because Iranian oil destined for the United States could have been sold to other buyers, and oil destined for those buyers could have come to the United States. However, the U.S. embargo on Iran was accompanied by very visible actions of the U.S. State Department against countries that entertained notions of purchasing the Iranian production made surplus by the U.S. action. In fact, in December there was a nasty confrontation between Japan and the United States in which the United States accused Japan of buying the additional supply, an action denied by Japan. After that episode, Iranian crude went to the spot market, where it was less visible.

The effect of the U.S. embargo was to slightly reduce the world supply of crude oil. November production dropped 200,000 barrels a day from October, and December production dropped another 500,000 barrels a day. Simultaneously, the spot value of Arab light increased from $33.27 to $37.09 a barrel between October and November and then increased further to $37.61 a barrel in December. These increases were relatively small, probably because the cut in production was not sharp, and by November 1979 inventories had been restored to higher levels. In fact, episode 5, like episode 2 which occurred only a year earlier, was remarkable for the small increases in prices. This causes us to wonder if Iranian crude was sold despite the embargo.

The final episode, episode 6, began in late September 1980 when war broke out between Iran and Iraq. This episode is unique because a sharp sudden cut in production was not accompanied by a large increase in the value of Arab light. The effect of the war on oil supply was dramatic. Over a sixty-day period free world production dropped by 3.6 million barrels a day from 59.5 to 55.9 million barrels a day—a cut matched only by the drop in 1973. The effect on prices was, however, less than might have been expected. The spot value of Arab light increased by 25.8 percent during the first month from $30.38 to $38.23 a barrel. By the end of October the price had dropped off to $35.84 a barrel.

As episodes 1 and 6 are compared, it is clear that there were many similarities and only one or two differences. The similarities were the size of the loss in production, the suddenness of the loss, the uncertainty of the duration, and the impact on specific purchasers who had long-standing contracts to buy from the countries making or experiencing the loss in production. The differences were the existence of an international energy apparatus, the willingness of another country to make up for some of the lost production, and stocks. The role played by the international energy apparatus, particularly the International Energy Agency, during this crisis cannot be assessed here because information on the role it played is, for quite obvious reasons, kept confidential. However, based on its failure during the Iranian crisis, we would tend to discount its effectiveness. The Saudi increase in production obviously reduced the magnitude of the shortage. The most important factor, however, and the clearest difference between the 1973 and 1980 episodes, was the level of consumer stocks. Stocks were high in the fall of 1980, and oil companies probably welcomed the opportunity to work them off. In 1973, on the other hand, stocks were low.

This discussion of the six events suggests some normative conclusions about the effect of a crisis on the spot price of oil. The first is that a disruption is more likely to cause a sharp increase in spot prices if there is a sudden, well-delineated loss in production. Episodes 1, 3, and 6 were each sudden, and the effect on prices in each case was dramatic. On the other hand, the slowly developing crises, such as episodes 2 and 5, tend to be characterized by smaller price increases, even if the loss in supply is as great.

The second normative conclusion is that the willingness of producers to make up for a loss in production in some other country is very important. This effect may be noted by contrasting the behavior of markets in episode 1, the Arab embargo, when several producers deliberately reduced production, with the behavior of markets in episode 6, when the war between Iran and Iraq caused a loss of production that was offset by increased production from Saudi Arabia, Kuwait, and other countries.

The third normative conclusion is that inventories are important. Here, again, the contrast may be drawn between the behavior of markets in 1973 and 1980. In the first episode inventories were low, and prices rose by very large amounts. In the case of the Iran/Iraq conflict, on the other hand, stocks were large, and the increase in prices was small.

These conclusions are summarized on Table 2-3, which indicates the four characteristics of each episode. The characteristics are the inventory situation, the type of disruption, the offsetting response of other countries, and the increase in prices.

## MODELING THE MOVEMENT OF SPOT PRODUCT VALUES

The preceding section described the movements during six prior episodes of disruption in oil markets and led to the conclusion that there is a relationship between the change in spot values, the change in supply, and the level of inventories. This was described in Table 2-2, where the ratio of the percentage change in prices to the percentage change in supply was compared to the number of days of supply. In this section we try to describe these movements with a model. The task requires that short-term changes in prices on commodity markets be modeled as a function of the conditions of supply. This problem is difficult. It is not one that can be expected to meet with great success, since other attempts to model the move-

Table 2-3. Normative Observation on Six Episodes of Sudden Change in Oil Markets.

| Crisis | Precrisis Level of Inventory | Type of Disruption | Offsetting Production Response | Maximum Change in Price |
|---|---|---|---|---|
| Embargo | Low | Sudden | No | 188% |
| Iran I | Medium | Slow | Yes | 10% to 20% |
| Iran II | Low | Sudden | No | 64.5% |
| Iran III | Low | Uncertain | No | 23% |
| Iran IV | High | Slow | Yes | 15% |
| Iran/Iraq | High | Sudden | Yes | 25% |

ments of prices on far more stable commodity markets have not met with great success.

The only previous attempt to explain movements of spot values (or spot prices) during disruptions is found in Nordhaus (1980b: 368). He argues,

> The final element in the oil market is the mechanism that equilibrates demand and supply. In the oil model the contract market is separate from the spot market. The contract market, which has varied from 90 to 98 percent of the total oil market, represents normal trading channels. The remainder, the spot market, is a kind of buffer that absorbs—and more fundamentally, reflects—underlying supply and demand shocks. In periods of excess supply, there will be discounts from the OPEC price on the spot markets, while during a shortage (such as the winter of 1973-74 or the 1979-80 period) the spot market price lies well above the official price.

Nordhaus (1980: 368) tests a hypothesis described by the following relationship.

> The spot price equation measures the discount or premium that exists relative to the official OPEC price. It is extremely nonlinear; below a critical value (97 percent of capacity) it is assumed that the spot price sells at a fixed discount relative to the official price. Above the official value (or kink), however, the supply function turns essentially vertical. Estimating the equation for the 1972-1979 period yields

$$Spoil_t = Poil_t \times \begin{cases} .977 \text{ if } Util \leq .97 \\ .430 + .543 g(Util_t) \text{ if } Util \geq .97 \end{cases} \quad (2.1)$$

$R^2 = .993$; Durbin Watson = 2.2; Standard error = .60

where  $Spoil$ = spot price of oil,
$Poil$ = OPEC price of oil,
$Util$ = ratio of oil demand to oil capacity,
$g(Util) = 5000 (Util - .97)^2 + 1$.

Under the Nordhaus model, spot prices are equal to 97 percent of official prices when there is at least 3 percent excess capacity in the world, but equal to an amount which is well in excess of the official price of oil when supply is tight.

There are three problems with the approach employed by Nordhaus. The first results from his use of annual data. This is an error because price changes during a disruption occur quickly, often within a few weeks or a few months. In fact, the longest period of upward movement during any disruption occurred in episode 1 when commodity prices and spot values increased continuously for three months. The use of annual data thus obscures the movement of prices during a disruption. This type of misspecification in time aggregation biases empirical results.

A more troubling problem with the Nordhaus work, however, is the fact that he uses statistical approaches to estimate two parameters with only two independent observations. In particular, he derives the relationship

$$Spoil_t = Poil_t \times [.430 + 0.543g(Util_t)]$$
$$\phantom{Spoil_t = Poil_t \times [}(.06) \phantom{+} (.04)$$

based upon observations on the values of $Spoil_t$, $Poil_t$, and $Util$ in 1973 and 1979. Of course, stated in this fashion it immediately may be observed that his statistical results are totally invalid because he has estimated two parameters based on two independent observations, leaving him with no degrees of freedom. In his paper, the problem is glossed over by introducing seven other observations in the regression for the years when capacity utilization was less than 97 percent. Such adjustments are only cosmetic, not real. Nordhaus has basically estimated two equations, one for normal times when capacity is less than 97 percent, where he has seven observations and six degrees of freedom, and one for periods of disruption, where he has two observations and no degrees of freedom. The first equation is believable; the second must be ruled out.

The greatest problem with the Nordhaus approach, however, results from the specification of his price equation. He argues that capacity is the only important variable and apparently measures capacity on estimates of maximum sustainable productive capacity developed by the Central Intelligence Agency (CIA). In his model, prices should remain stable so long as world demand for oil is less than 97 percent of capacity.

The problem with this approach is that none of the six episodes of price escalation during the 1970s was triggered by the loss of producing capacity—at least not as defined by the CIA:

- In episode 1 the increase in prices was triggered by a reduction in production which was simultaneous with the emargo. The reduction in production reduced the rate of capacity utilization and drove prices up.

- In episode 2 the price increase was triggered by a strike of oil field workers in Iran, which caused output to be reduced, again reducing capacity utilization and causing prices to increase.
- In episode 3 the increase in prices was triggered by the Saudi Arabian announcement that first quarter 1979 production would average 8.6 million barrels a day, thereby forcing a cut in production for several weeks (and simultaneously reducing the rate of capacity utilization).
- In episodes 4 and 5, price increases were triggered by actions that caused modest reductions in production—another way of reducing the rate of capacity utilization.
- In episode 6 the price increase did result from the loss of logistical capacity due to the war between Iran and Iraq. This loss of pipeline capacity did not affect the world's oil productive capacity and thus did not affect capacity as apparently defined by Nordhaus. (Capacity utilization did not, in fact, decline, because Saudi Arabia increased production to offset the loss from Iran).

Unfortunately, the Nordhaus paper appears to be the only study of movements of prices on the oil commodity markets. In attempting to develop a model that describes movements in prices, therefore, we must turn to studies of the behavior of prices in other commodity markets, particularly wheat and copper.

The studies we chose to follow are those developed by Fisher (1970), Fisher, Cootner, and Bailey (1973), and Taylor (1979). Our approach is to use these studies to model the movement of prices by estimating the short-run demand function for petroleum. This will be accomplished in a fashion that is different from that applied to the study of the demand for most commodities. Specifically, we intend to partition the data into two categories: observations generated during periods of normal market conditions and those generated during periods of shortage.

During periods of normal market conditions, we assume that the demand function will be given by a standard specification

$$Q_t = F(Z), \qquad (2.2)$$

where $Q_t$ is the quantity consumed and $F(Z)$ represents the demand function. One of the determinants of demand, $Z$, will include prices.

During periods of disruption, we assume that a different demand function applies. It is given by

$$SV_t = H(X), \tag{2.3}$$

where $SV$ represents the spot value of petroleum products and $H(X)$ represents the characteristics of the demand function. This approach theorizes that during periods of stable markets prices remain relatively stable over short periods while the quantity of oil consumed adjusts as determinants of consumption change. On the other hand, during periods of disruption we assume that the quantity of petroleum available to be consumed is fixed while prices adjust to balance the forces of supply and demand. We also assume that the movement is observed in roughly equal movements of all prices so that the change can be captured by a single aggregate price.[2]

The estimation of the movement in spot values through the specification of a demand function requires that we make one critical assumption that will allow us to identify the demand curve.[3] During disruptions we assume that the supply curve is totally inelastic (i.e., vertical) at any moment of time, a characteristic described in Figures 2-2a and 2-2b.

As the disruption progresses, it is assumed that the inelastic supply curve shifts from one position to another, and it is through the shifting process that the demand curve can be identified. (See Figure 2-2a.) This assumption will not work during periods when excess supplies of oil are available. At those times the supply curve is essentially infinitely elastic at a price set by OPEC. Thus, when excess volumes of oil are available, changes in the quantity of oil sold would not permit estimation of the price elasticity of the demand curve. (See Figure 2-2b.)

This behavior may be modeled by making the market clearing price, $SV_t$, a function of the quantity available

$$SV_t = \alpha_o + \alpha_1 Q_t + \epsilon_t, \tag{2.3}$$

2. Note that ideally the changes in individual prices would be modeled jointly rather than in aggregate. However, such an analysis is beyond the scope of this study.

3. The identification problem in econometrics is discussed in all standard econometrics texts and is treated very thoroughly in Fisher (1970). The problem can be described by noting that at any moment of time we observe a pair of observations on price and quantity generated by the equilibrium of supply and demand. Without information as to the characteristics that differentiate the supply curve from the demand curve, however, it is impossible to use this information on price and quantity to estimate either a demand or a supply curve.

**Figure 2-2a.** Identification of the Demand Curve During a Disruption.

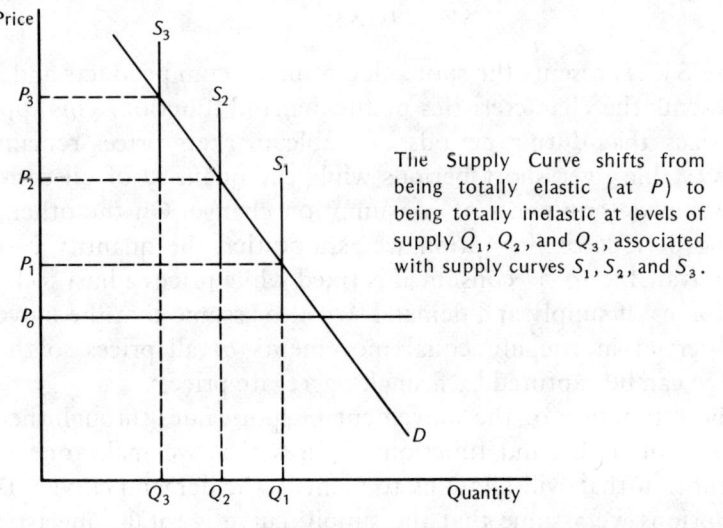

The Supply Curve shifts from being totally elastic (at $P$) to being totally inelastic at levels of supply $Q_1$, $Q_2$, and $Q_3$, associated with supply curves $S_1$, $S_2$, and $S_3$.

**Figure 2-2b.** The Price Coefficient of the Demand Curve Cannot Be Identified During Periods of Stable Supply at Constant Prices.

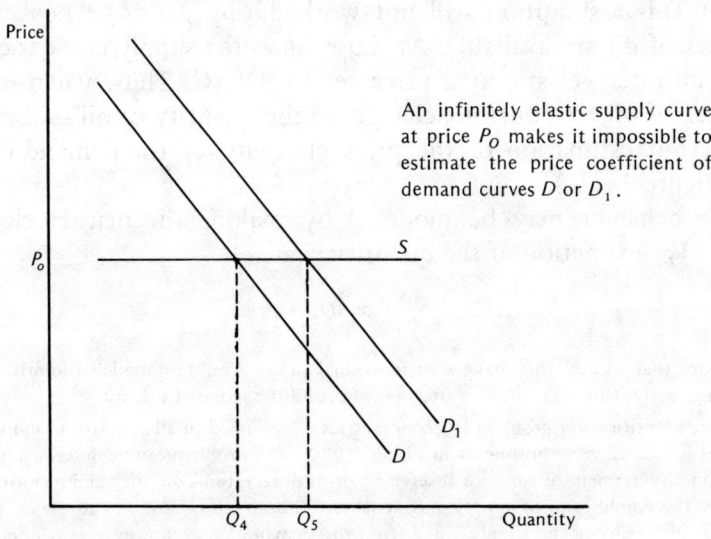

An infinitely elastic supply curve at price $P_O$ makes it impossible to estimate the price coefficient of demand curves $D$ or $D_1$.

where $Q_t$ represents supply, or, more precisely, world production of crude oil and $\epsilon_t$ an error term. The sign on the parameter $\alpha_1$ would be expected to be negative on the theory that a reduction in overall supply would cause prices to increase. Equation (2.3) may be generalized to include other variables, such as inventories ($K_t$) and income ($Y_t$):

$$SV_t = \alpha_0 + \alpha_1 Q_t + \alpha_2 K_t + \alpha_3 Y_t + \epsilon_t \ . \qquad (2.4)$$

Here one would expect to find $\alpha_2$ negative and $\alpha_3$ positive so that an increase in income would, ceteris paribus, cause an increase in price, while an increase in $K$ would cause a decrease in price.

The demand equations specified in (2.3) and (2.4) are assumed to apply at all periods of time. However, they are estimable only when the quantity supplied is fixed, and less than the quantity that would clear the market during a period of market stability. This assumption permits the demand curve to be identified.[4]

The demand equations specified in (2.3) and (2.4) also are assumed to describe changes in the relevant measure of price at which crude oil and products may be purchased during a disruption. This assumption will trouble some, particularly those such as the European Economic Community (EEC) officials who argue that the spot market is irrelevant. The contrary view is taken here: the spot market represents the arena where a willing buyer may acquire additional supplies of petroleum products at any time. Furthermore, to support this assumption, it is noted that during the 1979 crisis in the United States, many large users, such as automobile rental companies, turned to the spot market to acquire the products required to keep their operations going. The spot market is, in short, the incremental source of supply.

---

4. The point is trivial but must be made. Suppose, for instance, that in period 1 supply and demand are equal on the horizontal portion of the supply curve at, say $Q_0$. Assume further that initially an increase in demand would be accommodated by an increase in supply to $Q^*$ without any change in price, so that the kink in the demand curve occurs at $Q^*$. This implies that $SV_0 = H(Q_0) = H(Q^*)$. Under these circumstances, a disruption would have to involve the loss of at least $Q^* - Q_0$ in productive capacity in order to perturb $SV$. It also would take a cut of at least $Q^* - Q_0$ in supply in order for the demand equation to be estimated.

## ESTIMATION OF THE SPOT PRODUCT VALUE EQUATION

The demand equations such as those specified in (2.3) and (2.4) were fitted to monthly data on spot values, inventories, and measures of income generated during each of the six disruptions identified earlier in this chapter. The initial results were poor. The values of the estimated parameters were frequently statistically insignificant, and many had the wrong sign. According to the initial results, a reduction in the quantity of oil supplied did not necessarily induce an increase in spot values.

An examination of the residuals from the initial regressions and a study of the underlying data (which was drawn basically from Table 2-2) indicates that the changes in prices did not correspond exactly with the quarter-to-quarter change in production. At times the change in price would occur coincidently with the change in production, and at times the change in price would follow a month or two behind the change in production.

Not surprisingly, changes in price occurred coincidently with changes in the supply of oil when the change in supply was announced, as it was during the embargo (episode 1) and during the Saudi cutback in episode 2. In the other cases, where there was no announcement, the reaction in prices appeared delayed. This behavior should not seem surprising, since a change in supply may not be felt immediately in the marketplace due to the time required (six weeks) to transport oil from the Persian Gulf and Africa to the market.

One way of correcting for the differences in observed response would have been to introduce dummy variables for those crises where the cutback was announced and where the cutback was imposed in a relatively dramatic fashion. However, in lieu of using dummy variables, another approach was tried. In this latter approach, levels of expected production were calculated and the production shortfall was calculated under the hypothesis that price increases are induced when production falls short of expected levels of output. This relationship is expressed by

$$SV_t = \alpha_o + \alpha_1 (Q_t - Q_t^e) \; , \tag{2.5}$$

where $Q_t^e$ represents expected production.

The data on expected production were calculated by computing the seasonal pattern for world crude output and then computing the expected level of production for any period of disruption as a function of production during the last predisruption month and the relative seasonal factors. Thus, if $Q_o$ represents actual production in the month prior to the disruption, $S_o$ represents the seasonal factor for the month prior to the disruption month, and $S_t$ represents the seasonal factor for the month in question, then expected production in month $t$ would be written as

$$Q_t^e = Q^o \cdot (S_t/S_o) .^5 \qquad (2.6)$$

The values of $S_t$ were computed using simple averages of annual production for three years when there were no disruptions.

The computation of expected production takes into account the fact that worldwide oil production tends to fluctuate through the year with consumption, rising in the late summer and through the fall in anticipation of increased consumption during winter months and then declining in January or February in anticipation of declining rates of consumption during the spring and summer.[6] Data on the seasonal pattern of production are shown on Table 2-4.

In addition to introducing expected production, a second change was made in estimation of the price equation involving the measurement of the price variable. In the first analysis, month-to-month

---

5. Expected production may be characterized as the amount of supply purchasers of crude oil believe will be required to satisfy current consumption plus meet the desired level of stock change. Thus, letting $\epsilon(Q_t^D)$ represent expected demand and $\epsilon(\Delta I_t)$ represent the expected change in stock, expected production would be given by

$$Q_t^\epsilon = \epsilon(Q_t^D) + \epsilon(\Delta I_t). \qquad (2.6')$$

Unfortunately, data on refiner or producer expected consumption and expected changes in the level of stocks are unavailable except on an annual basis. Therefore it has been assumed that purchasers of crude oil normally would expect the pattern of production to follow a regular seasonal pattern. This means that month-to-month changes in production would be expected to reflect differences in seasonal adjustment parameters.

6. Some, such as Danielson (1979), have argued that this type of fluctuation is also the result of anticipation of OPEC meetings, which traditionally take place in December. However, a review of the data suggests that the pattern is observable both in years when OPEC met in December and in years when OPEC met at other times. A more careful examination of the data suggests that the timing of the decline in production in the spring is more dependent on the severity of the winter than OPEC decisions. During unseasonably warm winters, production declines early. During unexpectedly harsh winters, production remains high into the spring so that inventories can be rebuilt.

Table 2-4. Monthly Pattern in World Crude Oil Production.

| Month | Production as a Percentage of Annual Average |
|---|---|
| January | 93.6 |
| February | 96.9 |
| March | 97.6 |
| April | 98.7 |
| May | 97.6 |
| June | 98.9 |
| July | 98.4 |
| August | 99.1 |
| September | 100.6 |
| October | 102.6 |
| November | 103.6 |
| December | 104.7 |

changes in prices were computed from the peak price during the crisis. (These are the data shown on Table 2-2.) Thus, the spot value for month $t$ in any regression represented the spot value on the particular day corresponding to the day in the month in which the particular crisis was deemed to have peaked. (For instance, if the crisis caused prices to peak on October 17, the prices would be computed on November 17, December 17, etc.) These prices were replaced by a more traditional measure of prices—monthly averages.

The initial specification used in the final estimation of the price equation was

$$SV_t = \alpha_o + \alpha_1 (Q_t - Q_t^e) + \alpha_2 SV_{t-1} + u_t, \qquad (2.7)$$

where $SV$, $Q_t$, and $Q_t^e$ have been defined above and $u_t$ is an error term.

In equation (2.7) the level of spot values is specified as a function of the shortfall between contemporaneous production ($Q_t$) and expected production ($Q_t^e$), and lagged spot values.

It should be noted that an inventory variable was omitted from the final equation. There are two reasons for this. First, attempts to include an inventory variable failed.[7] Second, in the discussion in the

---

7. Both a measure of total stocks and consumption of stocks and a modification of $Q_t$ to reflect the change in inventories were attempted. None of the approaches worked. In

previous chapter we argued that movements in inventories were a lagged response to changes in the spread between spot values and official crude prices. Thus, the model used here assumes inventories respond to, but do not affect, movements in spot values. Essentially this means that inventories are held as a buffer to meet seasonal fluctuations and for speculative purposes. However, the results offered here also suggest that marketers and refiners intend to meet consumer demands from production.

A third reason why inventories proved to be a poor predictor of spot values relates to their relationship to output. Specifically, if stocks are thought to be low, then production prior to the crisis, $Q_o$, would be greater than if stocks were perceived to be high. Under this assumption, the level of stocks affects $Q_o$ but not the gap between $Q_t$ and $Q_t^e$.

The equations describing the behavior of spot market prices were estimated in both linear and logarithmic form. The logarithmic specification is shown in equation (2.7'):

$$\text{Log}(SV_t) = \alpha_o' + \alpha_1 \text{Log}(Q_e/Q_t) + \alpha_2 \text{Log}(SV_{t-1}) + u_t. \quad (2.7')$$

The statistical results from the estimation of (2.7) and (2.7') are shown on Table 2-5. The results are noteworthy because the sign on the term measuring the difference between expected and actual production is correct and because the parameters are statistically significant. In the linear equation the results indicate that a shortfall of 1 million barrels a day will cause spot values to increase by $0.41 in the first month and by $4.00 in the long run (assuming that expected production remains permanently greater than actual production). The logarithmic specification indicates that a 1 percent shortfall in actual production (500,000 barrels a day) would induce an immediate 1.5 percent increase in the spot price and a 9.0 percent increase in the long run (again assuming that expected production always exceeded actual production).[8]

---

part this failure may result because the inventory data is not available at a sufficiently high frequency to be useful.

8. Long-run impacts are derived by dividing the estimated parameter by one minus the coefficient of the lagged dependent variable. In assessing the long-run impact of a cut in production, we should recognize that equations (2.7) and (2.7') describe a permanent situation where expected production always falls short of anticipated production. Such an assumption is, of course, unreasonable, and thus the long-run calculations are really unrealistic.

Table 2-5. Results from the Econometric Estimation of the Adjustment Equation.

| Dependent Variable | Constant | Difference Between Expected and Actual Production | Lagged Spot Values | $\bar{R}^2$ | SEE |
|---|---|---|---|---|---|
| SV | 2.797 (1.404) | $-.410^a$ (.214) | $.916^a$ (.051) | .93 | 2.59 |
| SV |  | $-.522^a$ (.219) | $1.007^a$ (.023) | .92 | 2.75 |
|  |  | Log of Ratio of Expected to Actual Production | Log of Lagged Spot Values |  |  |
| Log (SV) | .460 (.150) | $1.495^a$ (.591) | $.864^a$ (.047) | .93 | .12 |
| Log (SV) |  | $1.880^a$ (.676) | $1.006^a$ (.010) | .91 | .14 |

a. Parameters statistically different from zero at the 99 percent confidence level. Standard errors in parentheses.

Results for equations (2.7) and (2.7') are also shown where the constant has been suppressed. The specification of these equations would be given by

$$SV_t = \alpha_1 (Q_t - Q_t^e) + \alpha_2 SV_{t-1} + u_t \tag{2.8}$$

and

$$\text{Log}(SV_t) = \alpha_1' \text{Log}(Q_t^e/Q_t) + \alpha_2' \text{Log}(SV_t) + u_t. \tag{2.8'}$$

The assumption underlying these specifications is that spot values change only when a gap between expected and actual production emerges.

The results, shown on Table 2-5, indicate that this specification fits almost as well as the specification including the constant. More significant, however, are the signs on the lagged dependent variables. These parameters are equal to unity in both the linear and the logarithmic equations. The implication of this finding is, of course, that

**Table 2-6.** Results from the Econometric Estimation of the Spot Value of Adjustment Equation Estimated in First Difference Form.

$$SV - SV_{t-1} = -.554^a \ (Q_t - Q_E)$$
$$(.193)$$

$\bar{R}^2 = .0932$ \hfill SEE = 2.702

$$\text{Log}\,(SV/SV_{-1}) = 2.06^a \, \text{Log}\,(Q_E/Q_t)$$

$\bar{R}^2 = .177$ \hfill SEE = 0.149

---

a. Parameter statistically different from zero at the 99 percent confidence level. Standard errors in parentheses.

the adjustment in spot values to any shortfall in production is instantaneous and is fully completed within one month. Such behavior can be modeled by rewriting equations (2.7) and (2.7') as first difference equations, as shown in equations (2.9) and (2.9'):

$$SV_t - SV_{t-1} = \alpha_1 (Q_t - Q_t^e) + u_t \qquad (2.9)$$

$$\text{Log}\,(SV_t/SV_{t-1}) = \alpha_1' \, \text{Log}\,(Q_t^e/Q_t) + u_t \ . \qquad (2.9')$$

The results from the estimation of (2.9) and (2.9') are shown on Table 2-6. As with the statistical results shown on Table 2-4, the parameter of the difference between actual and expected production is statistically significant. The measure of fit, the $\bar{R}^2$, has declined. However, this decline is illusory because the coefficient of multiple correlation measures the ratio of the explained to the unexplained variance in the dependent variable. In this case there has been little change in the unexplained variance. There has been a decline in the variance of the dependent variable, which is confirmed by the fact that the standard errors of the regressions shown on Table 2-5 are little different from the standard errors shown on Table 2-6.

The materials shown on Tables 2-5 and 2-6 lead to the following conclusions. First, a shortfall in actual production relative to expected production will induce an increase in spot values. Second, the increase will occur quickly, perhaps almost instantaneously. Third, for a shortfall of 1 million barrels a day (a reduction on the order of 2 percent) the price increase would be $0.55 a barrel in the case of the linear equation and approximately 4 percent in the case of the logarithmic equation, an increase that could be as much as $1.50 a barrel at today's prices.

The materials that follow adopt the logarithmic specification shown on Table 2-6 for two reasons: First, the fit is slightly better, and second, the logarithmic specification is scale free. Thus, a given percentage cutback in supply will have the same percentage effect on prices whether the price is $15.00 or $45.00 a barrel. Given the wild fluctuations in the oil market, this is thought to be a virtue.

As a final note on the estimation of the price equation, it should be pointed out that an examination of the error terms from each of the equations showed an absence of heteroskedasticity.[9] This finding is particularly important because equation (2.9) was fitted to a sample of twenty-five observations drawn from the six separate episodes from 1973 to 1979. Given the wide variation in prices, one might expect to observe a nonrandom pattern in the error terms. In particular, one might expect to observe heteroskedasticity since prices in 1973 were much lower than prices at the end of 1980.

## CONCLUSION

This chapter has examined the behavior of the spot value of crude oil during disruptions in oil markets. The discussion began by noting that oil markets had suffered through six disruptions during the 1970s, not three as is commonly believed. It was noted that the principal linkage between the commodity markets and the disruptions was supply as measured by the production of crude. When supply was unexpectedly reduced, spot values increased. It then was demonstrated that this relationship can be modeled, although not precisely.

The most important conclusion of the chapter, however, concerns the role of inventories. According to both the simple graphical analysis of the behavior of prices in the six episodes of market disruption and the econometric analysis of the six episodes, the level of stocks at the start of a disruption is important only to the extent that the stocks are used. This conclusion should not be surprising, because, as

---

9. "Heteroskedasticity" is a property used to describe errors from regressions. In the equation $Y_t = a + bX_t + u_t$, the error term $u_t$ has an expected value of zero and variance of $\alpha^2$. The residuals from the regression are said to be heteroskedastic if they appear to be proportional to $Y_t$, so that their average value increases.

noted in Chapter 5, inventories are acquired when the spread between spot and contract prices increases. Thus, whatever the level of stocks, there will be an inclination to change the rate of acquisition at the start of a disruption. If stocks are being liquidated, the rate of liquidation will be slowed or stopped. If they are being acquired, the rate of acquisition will be quickened. If these inclinations are followed, then the level of stocks will not tend to depress price increases, and under certain circumstances it could heighten them.

The conclusion should also not seem surprising since inventories represent only a small fraction of annual consumption and some portion of those stocks must be maintained to permit the smooth functioning of the refining process. This would mean that inventories would not be available to meet losses of supply resulting from emergencies.

This brings us to the essential conclusion of this chapter—that reductions in supply will require relatively quick compensating adjustments in consumption. During the six episodes examined, the reductions in consumption were accomplished by increases in prices. Given the reduction in supply, the simple model predicted fairly accurately the increase in prices that followed during each of the major disruptions in the 1970s.

# 3 THE RELATIONSHIPS BETWEEN THE COMMODITY MARKETS AND ADJUSTMENTS IN OFFICIAL OPEC PRICES

> It is my crude and I will do with it as I please.
>
> Attributed to Sheik Yamani at the May 25, 1981, meeting of OPEC in Geneva by *The Wall Street Journal* (May 27, 1981: 3).

Sheik Yamani's statement reflects the prevailing view of how official world crude prices are determined, that is, they are set by OPEC. The same view is expressed by another *Journal* report (May 13, 1981: 3), which observes, "Saudi Light currently costs $32.00 a barrel, but sources say Gulf expects that price to go up following the Organization of Petroleum Exporting Countries meeting in Geneva May 25. Gulf believes the price will rise to either $34.00 a barrel or $36.00 a barrel."

Of course, OPEC did not increase the price of crude at the May 25th meeting. Instead, general confusion arose after the meeting because Saudi Arabia insisted on holding down prices and used its surplus productive capacity to enforce its will on the twelve other members of the cartel who wanted to raise prices.

The inability of oil company officials to anticipate the outcome of the May 1981 meetings is surprising, especially in light of prevailing conditions on world oil markets through most of 1981. For instance, *Platt's Oil Price Service* published weekly estimates of the netbacks (see the Appendix) of various types of crude oil throughout the period that showed a steady deterioration in the value of all types of crude oil. Indeed, *Platt's* went so far as to publish a "buyer's balance" that indicated the profit or loss on a barrel of crude purchased at spot or contract for the period. These estimates (which are summarized on Table 3-1) showed that firms purchasing crude under contract were taking large losses throughout much of the period, a fact that undoubtedly was known all too well by the officials of Gulf Oil and every other company purchasing high-priced oil.

One possible reason for the oil companies' expectation that OPEC would increase prices at the May 25th meeting is that they believed the price of oil was determined by fiat rather than by competitive market conditions. This view was expressed by the President's Council of Economic Advisers (1978: 182) when it blamed the 1973 price increase on OPEC, stating, "OPEC quadrupled the world oil price levels far in excess of its cost and cut back production to support its action," and by Eckstein (1978: 113) who describes OPEC as "setting the price of oil." Many other references in recent literature express this view that "OPEC sets oil prices."

In fact, this view is an oversimplification of the process by which oil prices are set. Specifically, the periodic meetings of the cartel are held to validate price increases rather than announce them. Members of the cartel follow changes in product prices on the spot market (referred to here as "spot prices"). Before embarking on a technical explanation of OPEC pricing actions, we will state a general hypothesis about the mechanism by which OPEC controls prices.

Simply stated, that mechanism is supply. As Adelman (1979: 1) has noted, "Control of supply is the essence of monopoly; price fixing the result." In the case of oil, setting the level of production is the mechanism by which OPEC causes spot prices to increase. Thus, the members of OPEC contribute to the setting of the price of crude oil by their individual actions concerning supply. If supply is increased beyond current needs, spot prices are driven down and declines in prices are induced; when output is reduced, spot prices increase, permiting the members of the cartel to gather together to complain that ravenous and insatiable consumption in consumer countries has forced them to raise contract crude oil prices. This section adds statistical evidence to further support this view. It shows that market clearing prices that are reported to prevail for petroleum products on the principal petroleum commodity markets are the primary determinants of changes in official crude prices and

Table 3-1. Netbacks on Various Types of Crude Oil During the First Months of 1981 as Published in Platt's.

| | December 29, 1980 | January 29, 1981 | February 26, 1981 | March 26, 1981 | April 24, 1981 | May 14, 1981 | May 21, 1981 |
|---|---|---|---|---|---|---|---|
| | | | Dollars per Barrel | | | | |
| **Arab Light** | | | | | | | |
| Official price | $32.00 | $32.00 | $32.00 | $32.00 | $32.00 | $32.00 | $32.00 |
| Netback at source | 34.32 | 34.81 | 35.08 | 34.84 | 32.76 | 30.20 | 30.05 |
| Buyers balance | 2.32 | 2.81 | 3.08 | 2.84 | .76 | -1.80 | -1.95 |
| **Kuwait** | | | | | | | |
| Official price | 31.50 | 35.50 | 35.50 | 35.50 | 35.50 | 35.50 | 35.50 |
| Netback at source | 33.00 | 33.40 | 33.43 | 33.12 | 31.30 | 28.47 | 28.35 |
| Buyers balance | 1.50 | -2.10 | -2.07 | -2.38 | -4.20 | -7.03 | -7.15 |
| **Nigerian Light** | | | | | | | |
| Official price | 37.00 | 40.00 | 40.00 | 40.00 | 40.00 | 40.00 | 40.00 |
| Netback at source | 36.45 | 37.28 | 37.94 | 38.03 | 35.99 | 33.62 | 33.04 |
| Buyers balance | -.55 | -2.72 | -2.06 | -1.97 | -4.01 | -6.38 | -6.96 |

Source: *Platt's Oil Price Service.*

that a very systematic relationship between official and spot prices has prevailed since 1974.

The identification of the spot market as the principal determinant of crude oil prices is not new. In fact, the relationship has been understood among oil economists for years and has been described in some detail by Adelman (1972). More recently, Jacoby and Paddock (1980) have offered a description of the movement of official crude oil prices and spot prices that occurred during the chaotic market conditions of 1979. However, with the exception of Nordhaus (1980), no one has attempted to apply econometric techniques to the examination of the relationship between spot and official prices.

## THE THEORY

The hypothesis presented here is that crude oil, like any other unfinished commodity, is valued for the products derived from it. Thus, the spot value ($SV_{i,t}$) of a barrel of crude oil of type $i$ at time $t$ computed from the spot prices of products from the crude (see the Appendix) is thought to provide a measure of the value of crude oil. The assumption to be tested here is that the petroleum exporting countries use $SV_{i,t}$ to set the official price of crude oil, $P^c_{i,t}$.[1] However, it also is assumed that the process of price setting is not instantaneous, but one of continued adjustment—specifically, that the process is determined by a geometric lag process,

---

1. Note that a very strong assumption is being made as to the direction of causality between spot prices and contract prices. The assumption is based upon the behavior of the oil market in the 1970s and particularly four events. These occurred in October 1973 (the Arab/Israeli War), January 1979 (the collapse of the government in Iran), May 1979 (in response to regulatory actions in the United States and announcements by Saudi Arabia), and in October 1980 (the start of the Iran/Iraq war). Associated with each event were rumors of reductions in world oil output (some of which were later proven to be correct) and very large increases in crude values over a short period of time. (In 1973 the values increased by more than 180 percent from precrisis levels, while in 1979 values increased by 108 percent from starting levels.) These increases were followed by increases in official crude prices but with a lag of between three and six months. Thus in 1979, for instance, values increased by 60 percent from the end of January to early February after Saudi Arabia announced a cut in production, but official prices did not begin to reflect the higher values until early April and did not fully reflect them until early 1981. This pattern was repeated in other crises. Thus, following Adelman (1972 and 1979), we assume that the direction of causality is from a change in the world oil supply/demand balance (which may be caused either by a cut in production or a sudden increase in consumption) to the spot market, and then, after a lag, from spot prices to crude postings.

$$P^c_{i,t} = W \sum_{k=1}^{\infty} \alpha^k SV_{i,t-k} \tag{3.1}$$

where $0 < \alpha < 1$. Equation (3.1) can be transformed to an estimable equation,

$$P^c_{i,t} = \alpha P^c_{i,t-1} + \beta SV_{i,t-1} \tag{3.2}$$

where $\beta = W\alpha$.

Equation (3.2) expresses a relationship between increases in spot product prices and official prices. In its most general form, (3.2) specifies that a permanent increase of \$1.00 in $SV$ will cause an increase of $W\alpha/(1-\alpha)$ in $P^c$.[2]

This process may be described as follows. Assume first that spot values were absolutely stable until period 1. (Such a situation would prevail if $SV_{i,t} = P^c_{i,t}$ and $\alpha = 1-\beta$.) Suppose, however, that spot values increase permanently by \$1.00 in period 1. Then the model would predict that $P^c_{i,2}$ would increase by an amount $\beta$ and that $P^c_{i,3}$ would be equal to $\alpha\beta + \beta$ above $P^c_{i,0}$, and so on until, in the limiting case, $P^c_{i,t}$ increased by an amount equal to the increase in $SV$, which in this case was \$1.00.

The process described here assumes implicitly that official prices do not overshoot. This may be described by using a more restrictive form of equation (3.2), where the term $W \sum_{k=1}^{\infty} \alpha^k$ is forced to sum to unity. Under this hypothesis, one would assume that $P^c_{i,t} = SV_{i,t}$ if $SV_{i,t} = SV_{i,t-k}$ for all values of $k$. (In other words, if the value of a barrel of oil never changed, one would expect spot prices to equal official prices.)[3]

---

2. This follows because $\sum_{k=1}^{\infty} \alpha^k = \frac{\alpha}{1-\alpha}$, so long as $0 < \alpha < 1$.

3. A conclusion that $W \sum_{k=1}^{\infty} \alpha^k \neq 1$ would contain certain implications about the relationship between spot product and crude oil markets. Assuming $P^c$ and $P^*$ were correctly measured, a value of $W \sum_{k=1}^{\infty} \alpha^k$ greater than 1 would imply that consumers of crude oil (in this case primarily the major integrated firms that produce the oil) were willing to pay a premium to producers for the privilege of buying crude on contract. On the other hand, a value of $W \sum_{k=1}^{\infty} \alpha^k$ less than unity would imply that consumers of crude manage to appropriate some of the rents rightfully belonging to the producer.

This more restrictive hypothesis may be tested by imposing the condition

$$W \sum_{k=1}^{\infty} \alpha^k = 1, \qquad (3.3)$$

which can be satisfied when $W = (1-\alpha)/\alpha$. If this condition is imposed, the estimating equation becomes

$$P^c_{i,t} = \alpha P^c_{i,t-1} + (1-\alpha)SV_{i,t-1} . \qquad (3.4)$$

This equation resembles the specification estimated by Nordhaus (1980), who postulates a partial adjustment process.[4]

The basic model described by equations (3.2) and (3.4) was estimated using two alternative estimation techniques: ordinary least squares (OLS) and instrumental variables (I.V.). The estimating equation was specified as

$$P^c_{i,t} = \alpha P^c_{i,t-1} + \gamma SV_{i,t-1} + v_t , \qquad (3.5)$$

where, initially, $v_t$ was assumed to be distributed with zero mean and variance $\sigma^2$.

In the ordinary least squares estimates, actual values of $P^c_{i,t-1}$ were used as explanatory variables, while in the two I.V. cases, fitted values of $P^c_{i,t-1}$, $\hat{P}^c_{i,t-1}$ were substituted for $P^c_{i,t-1}$. These were computed by regressing $P^c_{i,t}$ on successively lagged values of $SV_{i,t}$.[5]

The two basic hypotheses, described in equations (3.4) and (3.5), were examined. Under the first, the parameter $\gamma$ was unconstrained in the regression in order to test the hypothesis described in (3.4),

---

4. Nordhaus (1980) estimates a single equation for all types of crude:

$$P^c_t - P^c_{t-1} = a(P^*_{t-1} - P^c_{t-1})$$

or

$$P^c_t = aP^*_{t-1} + (1-a)P^c_{t-1} .$$

Obviously, if $\alpha = (1-a)$, we arrive at the same specification, but generated by the partial adjustment equation

$$(P^c_t - P^*_{t-1}) = \alpha(P^c_{t-1} - P^*_{t-1}) .$$

Note, however, that Nordhaus uses a different measure of $P^*$.

5. Leviatan (1963) shows that this approach will yield consistent estimates of the parameters of the equation.

where $\gamma = \beta = W\alpha$. This restriction provides a test of the proposition that $W \sum_{k=1}^{\infty} \alpha^k = 1$.

Since both hypotheses cannot be maintained simultaneously, the statistical test described in Chow (1960) and Fisher (1970) was used to choose between them. Under this test, the ratio formed by dividing the difference between the sum of squared residuals in the unrestricted hypothesis and the sum of squared residuals in the restricted hypothesis adjusted by degrees of freedom is distributed as $F$ with 1 and $t-2$ degrees of freedom. If the computed ratio falls below the critical value of $F$, the second, restricted, hypothesis will be accepted, while if the ratio exceeds the critical value, the unrestricted hypothesis will be accepted.

## EMPIRICAL TEST OF THE THEORY

Equation (3.5) was estimated using three different quarterly measures of official crude prices for an interval from 1975 to 1980.[6] The types of crude were Arab 34, a 34° crude typical of the crude exported by Saudi Arabia; Arab 31°, a heavier crude, which is exported by all of the Persian Gulf countries; and African light, a crude typical of the Nigerian and Algerian exports. The price data were developed using data from *Platt's Oil Price Service* and *Petroleum Intelligence Weekly*. Estimates of petroleum products prices on the Rotterdam market published in *Platt's* were used to construct data on the value of a barrel of each crude together with estimates of the typical yield for each (used in forming the $w$'s) published in *PIW*.

The results shown on Table 3-2 provide estimates of the statistically estimated parameters $\alpha$ and $\beta$, as well as the calculated values of $W$ and $W \sum_{k=1}^{\infty} \alpha^k$, the ultimate effect of a permanent $1.00 a barrel increase in $SV_{i,t}$. The usual regression statistics are also given.

The following implications may be drawn from these results. First, the equation appears to fit very well to all three types of crudes. Second, a $1.00 per barrel increase in the value of products eventually induces more than a $1.00 increase in official prices. Third, the in-

---

6. The period of estimation was limited by problems of data definition. Prior to 1975 published reports of "official prices" have a different meaning. Thus, use of these data would be incorrect. See the Appendix for further details.

Table 3-2. Estimates of the Parameters for the Unconstrained Price Equation.

| Estimation Technique | α | β | $\bar{R}^2$ | DW | SEE | SSR | W | $W \sum_{k=1}^{\infty} \alpha^k$ [b] |
|---|---|---|---|---|---|---|---|---|
| | | | Arab Light | | | | | |
| OLS | .757[a] (.025) | .256[a] (.020) | .996 | 2.17 | .39 | 3.26 | .339 | 1.049 |
| IV | .784[a] (.039) | .228[a] (.031) | .991 | .69 | .58 | 7.12 | .291 | 1.051 |
| | | | Arab Heavy | | | | | |
| OLS | .667[a] (.039) | .381[a] (.036) | .994 | 2.04 | .51 | 5.64 | .575 | 1.147 |
| IV | .694[a] (.049) | .350[a] (.045) | .991 | 1.79 | .61 | 7.79 | .504 | 1.144 |
| | | | African Light | | | | | |
| OLS | .695[a] (.053) | .336[a] (.046) | .984 | 2.27 | 1.04 | 22.85 | .483 | 1.102 |
| IV | .714[a] (.067) | .314[a] (.057) | .976 | 2.04 | 1.23 | 32.09 | .440 | 1.094 |

OLS = Ordinary Least Squares
IV = Instrumental Variables
Standard errors in parentheses.
a. Statistically significant at the 99 percent level.
b. Estimate of the cumulative effect of a $1.00 permanent increase in $P^*$. Estimation period, 1975 to 1980: 3.
Source: *Platt's Oil Price Service* and *Petroleum Intelligence Weekly* (1975-81).

crease in the price of Arab light crude per $1.00 increase in $SV_{i,t}$ is greater than for Arab heavy and African crudes.[7] Finally, the results suggest that the producers of Arab light (the Saudis) are slower to respond to a change in $SV$ than the producers of Arab heavy or African light. According to the results presented here, a $1.00 increase in $SV_{i,t}$ for Arab light in the current quarter causes only a $0.26 increase in the official price of Arab light the next quarter, while simi-

---

7. This result is somewhat surprising, since it is generally believed that the producers of Arab light crude, primarily Saudi Arabia, are price moderates and that the producers of African and heavy Mid-East crudes tend to seek higher prices.

Table 3-3. Estimates of the Parameters for the Constrained Price Equation.

| Estimation Technique | α | $\bar{R}^2$ | DW | SEE | SSR | 1 - α |
|---|---|---|---|---|---|---|
| | | Arab Light | | | | |
| OLS | .719[a] (.016) | .995 | 1.83 | .42 | 3.82 | .281 |
| IV | .748[a] (.024) | .991 | 1.48 | .59 | 7.57 | .252 |
| | | Arab Heavy | | | | |
| OLS | .542[a] (.058) | .981 | .67 | .87 | 16.98 | .458 |
| IV | .578[a] (.063) | .981 | .67 | .89 | 17.47 | .422 |
| | | African Light | | | | |
| OLS | .615[a] (.044) | .981 | 1.64 | 1.14 | 28.52 | .385 |
| IV | .639[a] (.052) | .976 | 1.61 | 1.29 | 36.57 | .361 |

OLS = Ordinary Least Squares
IV = Instrumental Variables.
Standard errors in parentheses.
Estimation period, 1975 to 1980: 3.
a. Statistically significant at the 99 percent level.

lar increases in $SV_{i,t}$ for Arab heavy or African light crudes would induce $0.38 and $0.34 increases, respectively.[8]

The second step in the analysis was to impose the restriction $W = (\alpha - 1)/\alpha$, in order to test statistically whether the passthrough of increases in spot prices to official prices really was greater than one for one. The statistical results from the imposition of the restriction are given on Table 3-3. Here we show the values of $\alpha$, $(1-\alpha)$ and the usual regression statistics for the same crude oil examined above.

8. The increase in the $P^c_{i,t}$ in the current quarter following a $1.00 increase in $SV_{i,t-1}$ is given by $W\alpha$.

Once again we note that the equation fits extremely well and that all of the estimated parameters are statistically different from zero at the 99.5 percent level of confidence.

The results from estimation of the restricted hypothesis indicate that producers of the Arab heavy crude respond more quickly to changes in the spot market, while producers of Arab light are the slowest to respond. While this general result matches that obtained with the unconstrained model, the degree of sensitivity to the spot market changes.

The test of the proposition that $W = \frac{1-\alpha}{\alpha}$ is performed by applying the "F" test described in Chow (1960) and Fisher (1970). The results are shown on Table 3-4. The proposition that $W = \frac{1-\alpha}{\alpha}$ is accepted for Arab light at all levels of statistical significance, but re-

Table 3-4. Test of Hypothesis $H_1$, That $W \sum_{k=1}^{\infty} \alpha^k = 1$, versus $H_0$, That $W \sum_{k=1}^{\infty} \alpha^k \neq 1$.

| Estimation Technique | Value of Ratio | Comment |
|---|---|---|
| | Arab Light | |
| OLS | 3.61 | Accept $H_1$: $W \sum_{k=1}^{\infty} \alpha^k = 1$ |
| IV | 1.33 | " " " |
| | Arab Heavy | |
| OLS | 42.22 | Accept $H_0$: $W \sum_{k=1}^{\infty} \alpha^k \neq 1$ |
| IV | 26.09 | " " " |
| | African Light | |
| OLS | 5.21 | Accept $H_1$ at 99% confidence, $H_0$ at 95% confidence |
| IV | 2.93 | Accept $H_1$ |

OLS = Ordinary Least Squares
IV = Instrumental Variables.

jected at all levels of statistical significance for Arab heavy. For African light the result is ambiguous.

This apparent dichotomy suggests that the adjustment in prices made by the producers of Arab heavy and possibly African light represents an overreaction to changes in the market, and, indeed, the data appear to bear this out. Specifically, it can be noted that during the estimation period the price of Arab heavy declines twice and the price of African light declines three times, while the price of Arab light never declines.

## SIMULATION RESULTS AND CONCLUSIONS ON PRICE DETERMINATION

To determine the suitability of this approach, each of the estimated equations was simulated using actual spot values but endogenizing the prediction of the prior quarter's official crude price, which makes all predictions of future OPEC prices a function of only the starting price of crude oil and a sequence of spot prices. The results are summarized on Table 3-5. As the simulation results demonstrate, the empirical model accurately predicted movements in official prices during the 1974 to 1980 period. These results, then, lead us to the following conclusions.

Table 3-5. Mean Simulation Errors for Estimates of the Constrained Price Equation.

| Estimation Technique | Crude | MEAE | Standard Error | Maximum Error | Date of Maximum Error |
|---|---|---|---|---|---|
| OLS | Arab light | -0.26 | $.51 | $1.28 | 79:3 |
| IV | Arab light | -0.50 | .53 | -1.57 | 80:3 |
| OLS | Arab heavy | -1.08 | .84 | -3.47 | 80:3 |
| IV | Arab heavy | -1.20 | .87 | -3.72 | 80:3 |
| OLS | African light | -0.65 | 1.30 | 2.94 | 79:4 |
| IV | African heavy | -0.90 | 1.20 | 3.23 | 80:3 |

OLS = Ordinary Least Squares
IV = Instrumental Variables.

Estimates are based upon an endogenized value of the lagged dependent variable using the parameters given on Table 3-3.

First, the official price of crude oil set by OPEC countries is determined from past spot prices of the products on major world petroleum product markets. The elaborately staged meetings of OPEC members are totally irrelevant to the process of price determination. Instead, it is control over supply that is critical to the determination of crude prices. As Adelman (1979:1) notes, "The events since October 1978 emphasize that concerted control over supply is primary. Quoting of prices in concert is only incidental, convenient but not necessary to obtain the higher prices and revenues."

Second, Professor Adelman is correct. The control over output is everything. Thus, the cartel, or any member of the cartel, can start a movement in prices by adjusting output.

Third, price-setting behavior differs substantially among the various members of OPEC; some respond more rapidly to changes in the value of crude oil than others. This suggests that attempts to model short-term price behavior in aggregate will probably be biased.

Fourth, the so-called price moderates within OPEC—the producers of Arab light—are not following moderate pricing strategies at all but only adjusting prices at a slower rate. According to this analysis, all producers are following the market.

Finally, consuming countries have the power to prevent future surges in the price of oil such as those experienced in 1973–74 and 1979 by promoting calm and orderly spot markets and preventing panic buying and selling. This hypothesis is examined in more detail below.

# 4 THE BEHAVIOR OF CONSUMER PRICES

> Exxon USA, effective December 18, increased all grades of motor gasoline by 6 cents a gallon. Distillates, No. 2 heating oil, diesel fuel and kerosene, were up by 3 cents a gallon. Moves are nationwide to all classes of trade.
> Exxon cited the recent round of crude oil price hikes, some retroactive to November 1, as the reason behind the boosts.
>
> *Platt's Oilgram Price Report* 57, no. 244 (December 19, 1979): 6.

Chapter 1 identified four features of the petroleum market that could destabilize the market at the time of a disruption: the lethargic response of the members of OPEC to changes in spot values, the manner in which consumer prices are set, the tendency to acquire speculative inventories during the early phases of a disruption, and the behavior of spot market prices during the early phases of a disruption. It also noted that a shortage was likely to develop if consumer prices tended to follow changes in official crude prices.

This chapter demonstrates that prices paid for petroleum products by consumers have tended to follow changes in the cost of crude oil in both the United States and Europe. One explanation for this behavior is governmental regulation of prices. The chapter is organized into four sections: The first section reviews elementary price theory and demonstrates its relevance to the oil industry; the second argues that regulation may have prevented the oil industry from adjusting prices; and the last two offer extensive materials on changes in consumer prices and costs of crude.

## THE RELEVANCE OF PRICE THEORY TO THE OIL INDUSTRY

Three basic principles form the foundation of microeconomic theory of competitive markets: First, market equilibrium will occur when prices are allowed to adjust, so that supply equals demand; second, the supply curve in a market is determined by the aggregation of the marginal cost curves of the many atomistic firms; and third, the operating unit with the highest marginal costs will determine the equilibrium price where marginal costs differ.

Shortages should not occur in those markets where the participants adhere strictly to marginal cost pricing, but may occur where participants are either prohibited from following or elect not to follow marginal cost pricing. For instance, Chapter 1 showed that the failure of suppliers to raise prices at the time of a shift in the supply curve could create shortages in a market that previously had been in equilibrium. That explanation is repeated here because it forms the basis of the analysis in this chapter.

The market considered here is in initial equilibrium of supply $S$ and demand $D$ at $P_o Q_o$ as shown in Figure 4-1. This equilibrium is disturbed by a disruption (a loss in oil production, a crop failure, or some other type of disaster, depending on the market), which causes the supply curve to shift to the left from $S$ to $S'$. This shift creates a situation where the original equilibrium level of supply, $Q_o$, can no longer be supplied for $P_o$. Instead, $Q_o$ can be made available only if prices rise to $P_2$. However, due to the nature of the demand curve, consumers would no longer purchase $Q_o$ units of the commodity in question if prices rose to $P_2$. When the market is allowed to adjust to the new conditions of supply, prices increase only to $P_1$, which is less than $P_2$, and the quantity sold decreases to $Q_1$.

Suppose, however, that the market is not allowed to adjust because producers are prevented or unwilling to increase prices. Then the quantity demanded by consumers remains unchanged at $Q_o$ while the price remains unchanged at $P_o$. Under the new conditions of sup-

**Figure 4-1.** Effect of a Disruption on Product Market Equilibrium.

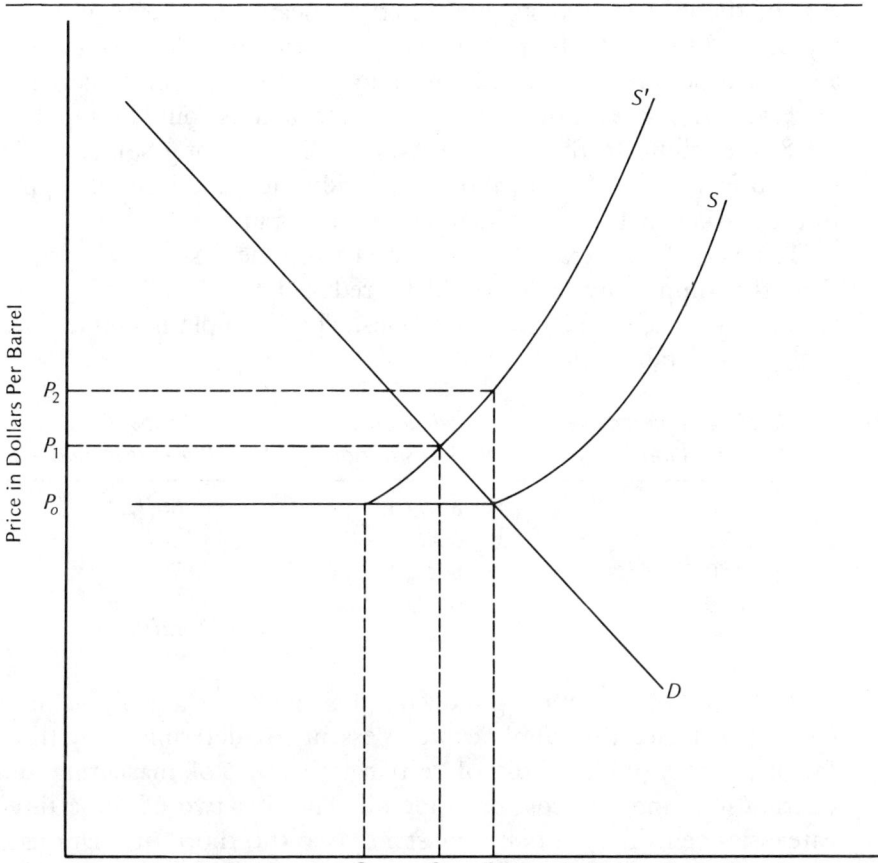

ply, however, producers are willing to supply only $Q_2$ units to the market at price $P_0$, leaving a shortage of $Q_0 - Q_2$.

To make this presentation more concrete, assume that the market described by Figure 4-1 is the gasoline market and that it has the following characteristics. Suppose that in the period prior to the disruption refiners can supply up to 100 billion gallons a year at 60¢ a gallon and that demand at 60¢ a gallon is 90 billion gallons. This would mean that $Q_0$ would be 90 and $P_0$ would be 60. Now assume that a disruption occurs so that refiners can only supply 70 billion gallons at 60¢ a gallon and that they could obtain further supplies of crude only at higher prices. Assume specifically that the higher cost of the incremental supplies of crude means that they could supply an additional 10 billion gallons at a price of 70¢ a gallon and additional supplies in 10 billion gallon increments for prices that increased by 10¢ a gallon, so that 90 billion gallons could be supplied for 80¢ a gallon. In these circumstances, the price of gasoline would have to increase to 80¢ a gallon to provide the same level of supply to the market that was provided before the disruption.

However, if refiners were prohibited from increasing their prices, then the supply available would be reduced to 70 billion gallons, leaving a shortage of 20 billion gallons. This example is summarized in the simple table below:

| Gasoline Supplied (billion gallons) | Marginal Cost Before Disruption | Marginal Cost After Disruption |
|---|---|---|
| 60 | 60¢/gal. | 60¢/gal. |
| 70 | 60¢/gal. | 60¢/gal. |
| 80 | 60¢/gal. | 70¢/gal. |
| 90 | 60¢/gal. | 80¢/gal. |
| 100 | 60¢/gal. | 90¢/gal. |

In the petroleum refining industry the shape of the marginal cost curve, and hence the supply curve, is essentially determined by three factors. These are the cost of refining, the cost of marketing and distribution, and the cost of crude oil. The first two of these three categories remain essentially constant over the short run. During a disruption, therefore, the change in marginal costs and hence the change in the shape of the supply curve will be determined almost entirely by changes in the marginal cost of crude. This means that the movement of the supply curve depicted in Figure 4-1 will be caused by changes in the cost of crude oil. This implies that changes in consumer prices of petroleum products (exclusive of tax) should reflect the fluctuations in the cost of the incremental supplies of crude oil during both times of glut and times of shortage. In other words, consumer prices should be predicated on the price of crude oil sold on the spot market.

This conclusion is based on the following considerations. The cost of incremental supplies should determine consumer prices. Since many refiners obtain their incremental supplies of crude oil from the spot market, spot costs should be the basis of the determination of prices. Finally, as noted above, the high-cost firm should set the price for all firms in a competitive industry.[1]

These conclusions as to the manner in which petroleum prices should be set are relevant to our assessment of the behavior of consumer prices of petroleum products during a disruption because the price at which incremental sources of crude oil may be obtained tends to follow closely the current level of spot product values. This behavior may be observed by a comparison of the fluctuations in spot product values with the fluctuations in spot crude prices. Such a comparison is shown on Table 4-1, from data drawn from *Petroleum Intelligence Weekly* on both the spot product values and spot crude prices.

The implication of this analysis is that the prices paid by consumers during disruptions should increase rapidly. This would cause the profits of many oil companies to rise very rapidly because they obtain much of their oil at the lower official prices, which tend to change very slowly in response to changes in spot values. These profits might have reached 36¢ a gallon during the Iranian disruption. (Normal profits are in the range of 1¢ to 2¢ a gallon.) Of course, refiners did not follow marginal cost pricing in either the 1973-74 or the 1979 disruption. One factor which may have prevented them from doing so was regulation.

---

1. This point may require some explanation. Where there are several firms in an industry, the industry supply curve is determined by the horizontal aggregation of marginal cost curves. Thus, if firm $i$ could supply $q_i$ units to the market at a price of $P$, industry supply at price $P_i$ is determined by summing $q_i$ over all firms. If

$$q_i = F_i(P) ,$$

and $\partial q_i / \partial P > 0$ (i.e., that the marginal cost curves slope upward), then industry supply is

$$Q = \sum_{i=1}^{n} q_i = \sum_{i=1}^{n} F_i(P)$$

and $\partial q / \partial P > 0$.

Now, if all firms but one have a maximum output $q_i^*$, at, say $P^*$, and market demand exceeds the aggregate output of these firms at $P^*$, then the one remaining firm that by assumption can increase output would be expected to increase output until the total output of all firms equalled demand at a price $P'$. $P'$ would be determined by the condition that demand at $P'$, which is written as $Q'$, equals the available supply from the last firm less the supply from the other firms. In other words, let

$$q_n' = Q' - \sum_{i=1}^{n-1} q_i^* .$$

Then, at equilibrium, $q_n' = F_n(P')$ for the nth firm.

**Table 4-1.** Comparison of Spot Product Values and Spot Crude Oil Prices of Mideast Light and African Light, 1978-80 (*dollars per barrel*).

|  | Mideast Light | | African Light | |
|---|---|---|---|---|
| Date | Spot Crude Price | Spot Product Value | Spot Crude Price | Spot Product Value |
| *1978* | | | | |
| January | $12.66 | $12.63 | $14.05 | $13.71 |
| February | 12.66 | 12.46 | 14.00 | 13.50 |
| March | 12.66 | 12.64 | 13.95 | 13.88 |
| April | 12.68 | 12.70 | 13.83 | 14.13 |
| May | 12.70 | 12.81 | 13.90 | 14.29 |
| June | 12.73 | 12.72 | 13.95 | 14.06 |
| July | 12.77 | 12.66 | 13.89 | 14.14 |
| August | 12.79 | 13.06 | 13.96 | 14.82 |
| September | 12.80 | 13.01 | 14.09 | 14.82 |
| October | 12.85 | 13.20 | 14.35 | 15.42 |
| November | 13.20 | 15.06 | 14.90 | 18.23 |
| December | 14.50 | 14.24 | 16.25 | 16.79 |
| *1979* | | | | |
| January | 15.95 | 16.33 | 18.95 | 19.71 |
| February | 19.50 | 22.21 | 22.00 | 27.08 |
| March | 20.80 | 22.13 | 23.25 | 25.49 |
| April | 21.20 | 23.61 | 23.65 | 27.65 |
| May | 34.25 | 27.82 | 32.75 | 33.52 |
| June | 32.85 | 31.76 | 36.50 | 37.67 |
| July | 32.00 | 30.61 | 35.00 | 34.35 |
| August | 32.25 | 29.89 | 35.75 | 33.18 |
| September | 34.50 | 30.23 | 36.50 | 34.09 |
| October | 36.00 | 31.74 | 38.50 | 35.79 |
| November | 39.50 | 34.65 | 42.00 | 39.45 |
| December | 39.00 | 35.68 | 40.50 | 40.49 |
| *1980* | | | | |
| January | 38.00 | 33.38 | 40.00 | 38.32 |
| February | 36.00 | 31.07 | 38.50 | 35.80 |
| March | 35.75 | 30.32 | 38.25 | 34.82 |
| April | 35.00 | 31.44 | 38.15 | 32.27 |
| May | 35.60 | 32.08 | 38.50 | 35.98 |

Table 4-1. continued

|  | Mideast Light | | African Light | |
| --- | --- | --- | --- | --- |
| Date | Spot Crude Price | Spot Product Value | Spot Crude Price | Spot Product Value |
| June | 36.00 | 30.90 | 38.00 | 34.83 |
| July | 33.35 | 30.32 | 37.40 | 34.10 |
| August | 32.30 | 28.72 | 33.60 | 31.58 |
| September | 32.25 | 28.92 | 33.40 | 31.58 |
| October | 36.80 | 32.90 | 37.90 | 34.25 |
| November | 39.75 | 36.10 | 40.85 | 37.18 |
| December | 39.35 | 34.55 | 40.15 | 34.97 |

Source: *Petroleum Intelligence Weekly* (special supplement Feb. 2, 1981) (see Appendix A).

## REGULATION OF PRICES

The U.S. oil industry continuously operated under one form or another of price controls from August 1971 to January 1981. From 1971 to the spring of 1974 these controls were imposed under the Economic Stabilization Act (ESA). Then from 1974 to 1981 mandatory controls were imposed under the Emergency Petroleum Allocation Act (EPAA). From October 1978 to January 1981 additional voluntary price controls were imposed by executive order of the president. The latter controls were administered by the Council on Wage and Price Stability (COWPS).

While the details of these controls varied over time, they essentially limited increases in the prices of petroleum products to increases in the average cost of purchased materials. The effect was first, to prevent firms from increasing prices when the cost of incremental supply increased and second, to prevent them from realizing the large increase in profits which would have been realized had they followed marginal cost pricing.

The Emergency Petroleum Allocation Act provides a good example of the type of controls imposed on the industry. The EPAA stipulated that changes in prices of petroleum products be limited to dollar-for-dollar passthrough of net increases in the cost of crude oil and imported products. Furthermore, it required that increases in the prices of specific products (heating oil, diesel fuel, aviation fuel, and propane) must represent no more than a direct proportion of the increased crude oil cost.

The effect of these regulations was to perpetuate the regulations established by the Cost of Living Council in August 1973. Ceiling prices were established for domestically produced oil, and a complicated transfer system (referred to as the entitlements system) was established to transfer costs between refiners so that all refiners paid the same marginal and average cost for crude oil (Kalt 1980). In addition, the EPAA required that price controls be imposed on petroleum products. These controls essentially required that the price of a product be based upon the average cost of crude oil plus a margin that was based upon an individual firm's margin in 1973.

The regulations (Section 212.83(c)(iii)(D), (E)) permitted refiners to recover increases in the average cost of crude oil and increases in nonproduct costs in the current period, or in subsequent periods if market conditions would not permit their recovery in the current period. The order of cost recovery, however, has been the subject of litigation and a number of changes in the regulations.

Through most of the period of regulation, Department of Energy (DOE) price ceilings were nonbinding for most firms because margins were frozen at all-time peaks. As Bohi and Russell (1978: 218) note, all refiners' product prices were decontrolled from January to March 1973 during a period of market tightness.

The COWPS ceiling price guidelines in effect from October 1978 to January 1981 established price standards with exceptions based on margins during two of the three years from October 1, 1975, to September 20, 1978. These regulations offered producers two alternatives: (1) They could limit their cumulative price increase to 0.5 percent below the firm's average range of price increase during 1976-77 or (2) if that was impossible due to increases in raw materials prices, they could satisfy a two-part profit margin test that restricts margins to the average margin for two of the last three years and not more than a 6.5 percent year-over-year increase (Council of Economic Advisers 1979: 82).

Because COWPS regulations were imposed at a time when margins were low, they were more stringent than the DOE regulations. (It should be noted that most oil companies were forced to follow the profit margin test because their raw material costs have increased at a rate much greater than the rate assumed in the original COWPS program.) The fact that unrecovered costs increased through 1979 and 1980 while the COWPS staff concluded in February 1980 that seven or eight of the twenty-five major oil refiners were slightly out of compliance with the program is evidence that it acted as a greater constraint than the DOE regulations (Council on Wage and Price Stability 1980).

These guidelines prevented consumer prices of petroleum products from adjusting rapidly when a disruption occurred. Refiners were required or coerced to hold their consumer prices at precrisis levels until their costs of crude increased, instead of rapidly increasing them as spot prices rose at the start of the disruption. Therefore, supply of products distributed to the market was reduced—following the predictions of our simple theoretical model.

## BEHAVIOR OF U.S. PRICES DURING THE IRANIAN DISRUPTION

This section examines the movements of consumer prices of various petroleum products in the United States during the Iranian disruption and demonstrates that prices changed with changes in the cost of crude oil. The analysis compares the cumulative increase in the spot value of crude oil since the start of the disruption with the cumulative increase in the official cost of crude and the cumulative increase in the price of product during the same period of time.

The data displayed on Table 4-2 show the cumulative increase in product and crude oil prices after June 1978. Table 4-2 also displays the cumulative increase in the prices of three categories of crude oil (measured in cents per gallon) subsequent to June 1978: a refiner composite, the Nigerian contract price, and the spot value of Nigerian light crude oil. The refiner composite represents the average price of crude paid by U.S. refiners for all categories of crude oil at a particular time, after adjustments for the effect of U.S. crude oil price controls. The African contract price is the price reported by *PIW* for African light crude oil. The spot value is that of African light crude (computed at Rotterdam).

From Table 4-2 it may be observed that the cumulative increase in the prices of diesel, gasoline, heating oil, and residual fuel oil dur-

**Table 4-2.** Cumulative Increase in Consumer Prices of Various Petroleum Products Since June 1978 in the United States as Compared to the Increase in the Cost of Crude *(cents per gallon)*.

| | Products | | | | | Crude | |
|---|---|---|---|---|---|---|---|
| Date | Diesel | Gasoline | Heating Oil | Residual Fuel Oil | Refinery Composite Price | African Light Contract | Spot Value African Light Crude |
| October 1978 | 0.9 | 1.1 | 1.7 | 0.4 | 0.5 | 0.0 | 3.0 |
| January 1979 | 3.0 | 3.4 | 5.2 | 3.0 | 1.7 | 2.0 | 13.3 |
| April 1979 | 10.6 | 11.9 | 12.6 | 9.0 | 5.0 | 10.8 | 32.2 |
| July 1979 | 28.9 | 26.9 | 25.3 | 16.8 | 14.7 | 22.6 | 48.1 |
| October 1979 | 34.8 | 34.5 | 33.8 | 20.8 | 19.7 | 24.5 | 51.5 |
| January 1980 | 42.2 | 44.9 | 42.3 | 31.8 | 29.5 | 45.7 | 57.6 |
| April 1980 | 48.0 | 58.1 | 48.9 | 23.9 | 35.0 | 51.8 | 43.2 |
| July 1980 | 47.6 | 58.6 | 49.4 | 26.2 | 38.9 | 57.7 | 47.5 |
| October 1980 | 45.7 | 56.2 | 50.2 | 29.6 | 40.9 | 54.8 | 47.9 |

Note: All price changes measured from June 1978.
Source: Department of Energy, *Monthly Energy Review* (various issues 1978–80).

ing the period of the Iranian crisis exceeded the cumulative increase in the refinery composite but fell far short of the cumulative increase in the spot value of African light. The cumulative increases in the prices of these products matched almost penny for penny the increases in the contract price of African crude. Thus, in July 1979, when the cumulative increase in African crude was 22.6¢ a gallon, cumulative increases in product prices were: diesel fuel, 28.9¢; gasoline, 26.9¢; heating oil, 25.3¢; and residual fuel oil, 16.8¢. Similar results are observed for other months. These data therefore tend to reinforce the assertion that changes in product prices tended to follow changes in the world price, not the spot value, of crude oil.

The comparison of changes in prices with changes in crude costs may be simplified by computing weighted averages of product price increases and comparing the single average with the increase in crude costs. Such a comparison is shown in Table 4-3. Column 1 shows the cumulative increase in the average consumer price of petroleum products computed from June 1978. Columns 2, 3, and 4 show the

Table 4-3. Cumulative Increase in U.S. Weighted Average Product Prices from June 1978 as Compared to Crude Oil Price Increases (*cents per gallon*).

| Date | Cumulative Increase in Weighted Average Product Price | Difference Between Cumulative Increase in Weighted Average Product Prices and Cumulative Increase in Crude Costs[a] | | |
|---|---|---|---|---|
| | | Refiner Composite | African Light Contract | Spot Value African Light |
| October 1978 | 1.0 | + 0.5 | +1.0 | - 2.0 |
| January 1979 | 3.6 | + 1.9 | -0.1 | - 9.7 |
| April 1979 | 11.3 | + 6.3 | +0.4 | -20.9 |
| July 1979 | 24.6 | + 9.9 | +2.0 | -23.5 |
| October 1979 | 31.4 | +11.7 | +6.9 | -20.1 |
| January 1980 | 41.4 | +11.9 | -4.3 | -16.2 |
| April 1980 | 48.2 | +13.2 | -3.6 | + 5.0 |
| July 1980 | 49.1 | +10.2 | -8.6 | + 1.6 |
| October 1980 | 48.5 | + 7.6 | -6.3 | + 0.6 |

a. The figures in the second through fourth columns represent the difference between the value in the first column and the cumulative increase in the cost of crude shown on Table 4-2. Weights reflecting a hydroskimming yield were used in computing the weighted average increase of product prices. (See the Appendix for further details.)

difference between the cumulative increase in average consumer prices and the cumulative increase in the three categories of crude prices that were tabulated on Table 4–2. Thus, from Table 4–3 it may be observed that average consumer prices had increased by 24.6¢ a gallon between June 1978 and July 1978 and that this increase was 9.9¢ a gallon *greater* than the increase in the refiner composite, 2.0¢ a gallon greater than the cumulative increase in the contract price of Nigerian crude, but 23.5¢ a gallon *less* than the cumulative increase in the spot value of Nigerian crude. In general, a study of columns 2, 3, and 4 should convince the reader that consumer prices moved most closely with the contract prices of African crude (and by implication most imported crudes, since the price charged for African crude is representative of the price charged for most imported crudes) and diverged sharply from movements in spot values and the regulated refiner composite.

This point is made even more clearly by Figure 4–2, where movements in U.S. consumer prices are compared to both the movement of spot product values for African light and the official price of African light. It may be observed that the consumer price (labeled "Normalized Value of Consumer Prices") moves closely with the official price of African light. (The normalized value series represents a weighted average of U.S. consumer prices, with the weights representing the typical product slate for a U.S. refinery that processes African light crude, with the average adjusted to equal the official price of African light crude in January 1978).[2] It also may be observed that the spot value of African light, computed from the Rotterdam product prices, diverges sharply from both the normalized price and the official price of African light. This, then, leads to the conclusion that U.S. consumer prices tended to follow changes in world prices of crude oil as set by petroleum exporting countries, as was suggested in Chapter 1.

---

2. The normalized price shown on Figure 4–2 represents a computation of refinery realizations using Gulf Coast refinery distillation weights for African crude and consumer prices for gasoline (3 grades), heating oil, and residual fuel oil. A constant sum of $8.72 a barrel was subtracted from every observation so that the January 1978 normalized prices would equal the January 1, 1978 official price of African light crude.

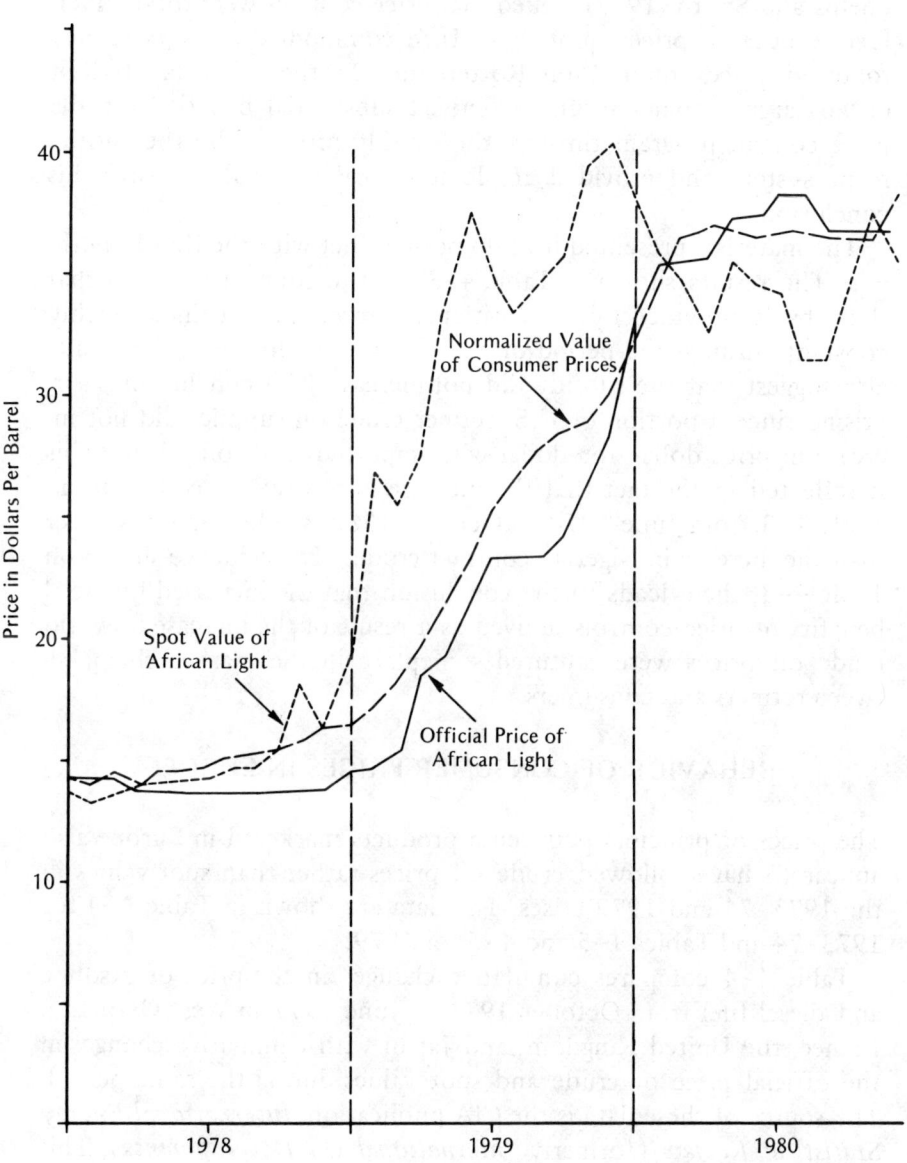

Figure 4-2. Comparison of Official Price of African Light, Spot Value of African Light, and Normalized Value of Consumer Product Prices in the United States.

## A Digression on Passthrough

Between 1977 and 1979 a number of economists prepared long treatises addressing the effectiveness of price controls. For instance, Phelps and Smith (1977) argued that price controls were totally ineffective because prices quoted on U.S. commodity markets closely followed prices quoted on Rotterdam. On the other hand, Kalt (1980) argued that consumers derived substantial benefit from the price control program through the subsidy provided by the entitlement system and provided ample statistical material to justify his conclusion.

The materials presented here do not conflict with the Kalt hypothesis. The results shown in Table 4-3 tend to support the view that the subsidy provided by the entitlements program remained roughly constant through the period of the disruption. However, the results also suggest that the subsidy did not increase. This conclusion is surprising since a portion of U.S. refiner crude oil supplies did not increase in price dollar-for-dollar with imported crude oil prices. (This is reflected in the fact that the increase in the refiner composite in Table 4-3 from June 1978 to October 1980 is substantially smaller than the increase in Nigerian contract crude.) The evidence shown on Table 4-3, then, leads to the conclusion that the increased financial benefits of price controls derived as a result of the increase in world crude oil prices were captured someplace in the market chain between refiners and consumers.

## BEHAVIOR OF CONSUMER PRICES IN EUROPE

The prices of principal petroleum products marketed in Europe also appear to have followed crude oil prices rather than spot values in the 1973-74 and 1979 crises. The data are shown in Table 4-4 for 1973-74 and Tables 4-5 and 4-6 for 1979.

Table 4-4 compares cumulative changes in the price of gasoline and diesel fuel from October 1973 to June 1977 in West Germany, France, the United Kingdom, and Japan with cumulative changes in the official price of crude and spot values during the same period. The source of these data is the CIA publication, *International Energy Statistical Review* (formerly *International Oil Developments*). This publication provides data on consumer prices of gasoline and diesel

Table 4-4. Cumulative Increase from October 1973 in Consumer Prices of Gasoline and Diesel in Various Countries During the Arab Embargo as Compared to the Price of Contract Price of Crude Oil and Spot Value of Crude Oil (*cents per gallon*).

| Month | West Germany | France | United Kingdom | Japan | Official Price | Spot Value |
|---|---|---|---|---|---|---|
| | | | Gasoline | | | |
| January 1974 | 15.3 | 18.0 | 2.9 | 24.2 | 16.4 | 26.0 |
| June 1974 | 20.5 | 19.4 | 24.6 | 37.1 | 18.3 | 16.0 |
| January 1975 | 14.9 | 28.4 | 56.9 | 46.1 | 18.9 | 14.9 |
| June 1975 | 15.4 | 34.1 | 54.1 | 51.0 | 18.1 | 17.5 |
| January 1976 | 21.8 | 30.1 | 35.7 | 48.5 | 21.5 | 17.7 |
| June 1976 | 25.1 | 26.5 | 27.5 | 50.8 | 22.0 | 19.3 |
| January 1977 | 29.4 | 25.6 | 29.9 | 52.5 | 25.0 | 21.8 |
| June 1977 | 31.4 | 31.3 | 26.2 | 58.7 | 25.0 | 21.5 |
| | | | Diesel | | | |
| January 1974 | 15.5 | 5.7 | 2.9 | 4.5 | 16.4 | 26.0 |
| June 1974 | 21.3 | 6.7 | 27.4 | 20.9 | 18.3 | 16.0 |
| January 1975 | 26.1 | 25.1 | 34.4 | 24.6 | 18.9 | 14.9 |
| June 1975 | 26.8 | 16.6 | 24.9 | 29.5 | 18.1 | 17.5 |
| January 1976 | 21.4 | 21.5 | 31.0 | 31.2 | 21.6 | 17.7 |
| June 1976 | 22.0 | 17.5 | 23.4 | 33.0 | 22.0 | 19.3 |
| January 1977 | 26.5 | 19.2 | 31.0 | 35.1 | 25.0 | 21.8 |
| June 1977 | 25.4 | 23.7 | 35.0 | 35.9 | 25.0 | 21.5 |

Note: Calculations exclude change in consumer tax.

Source: CIA *International Energy Statistical Review* and *Petroleum Intelligence Weekly*. (various issues 1974-77).

fuel inclusive and exclusive of taxes. The data shown on Table 4-4 are taken from various issues of the *International Energy Statistical Review* but were modified to correct for fluctuations in exchange rates.[3]

It may be observed from these data that increases in the prices of gasoline and diesel fuel in the 1973-74 period were less than the

3. The CIA publishes prices converted at the latest rate of exchange. For example, the price for gasoline in Germany in June 1975 may change even in 1981 as the exchange rate fluctuates. To make the calculations shown on Tables 3-4, 3-5, and 3-6, all prices were calculated using the original rate of exchange.

increase in even the official price of crude oil in West Germany, France, and the United Kingdom. While official crude oil prices increased by 16.4¢ a gallon from October 1973 (again using African light as our benchmark), gasoline prices increased by only 15.3¢ a gallon in Germany, 18¢ a gallon in France and 2.9¢ a gallon in the United Kingdom. (These calculations exclude any changes in taxes.) Increases in diesel prices were even smaller, so that the increase in the weighted average price of products probably would have been less. (The weighted average price cannot be computed because prices of heating oil and residual fuel oil are unavailable.) Note that the increase in spot values of crude oil was considerably larger than the increase in product prices. Apparently market prices in Europe did not follow changes in the spot market. Of course, as Kalt notes, an explanation for this behavior may lie in regulation. In 1973 and 1974 prices of petroleum products were regulated in France and the United Kingdom. In fact, only West Germany had a truly free market (Kalt 1980).

The existence of regulation probably explains the fact that the price of diesel fuel, a politically sensitive price in Europe, increased by almost the same amount as gasoline in West Germany (15.5¢ a gallon versus 15.3¢ a gallon) between October 1973 and January 1974, while increasing much less in France (5.7¢ a gallon), the United Kingdom (2.9¢ a gallon), and Japan (4.5¢ a gallon). It is also interesting to note that the unregulated German market appears to have followed official prices and not spot values in 1974.

Consumer gasoline prices in Japan apparently followed an entirely different pattern in 1974. Although these data exclude taxes, some governmental action that we have been unable to identify was apparently taken to raise prices. (After 1978 *The International Statistical Review* no longer carried data on consumer prices in Japan, apparently because the data were not comparable to those for the United States and Western Europe.)

### The 1978-79 Experience

Price changes during the Iranian crisis reveal a slightly different pattern of behavior. During the first six months of the crisis, prices of both gasoline and diesel fuel in West Germany increased almost penny-for-penny with spot values, while prices of the same products in France, Italy, and the United Kingdom increased at a much slower rate. However, by June 1979 cumulative increases in product prices in West Germany, France, and Italy were all keeping pace with increases in the contract price of crude rather than the spot price, while prices in the United Kingdom had increased by somewhat more than the contract crude prices.

In June official prices of crude oil had increased by only 16.9¢ a gallon (from June 1978), but spot values had increased by 56.2¢ a gallon. Gasoline prices had increased by 21.8¢ a gallon (from June 1978) in West Germany, 16¢ a gallon in France, 36¢ a gallon in the United Kingdom, and 2¢ a gallon in Italy. Then, over the next twelve months, gasoline and diesel prices in each of the countries increased by between 20¢ and 40¢ a gallon, while official prices increased by 30¢ a gallon and spot crude values *declined* by 6.8¢ a gallon. (See Tables 4-5 and 4-6.)

The data on consumer prices displayed on the three previous tables appears to support the assumption that product prices are changed in response to changes in the price of crude oil—that is, in the administered contract price set by the petroleum exporting countries—and not in the price that would be determined from the spot

Table 4-5. Cumulative Increase in Consumer Gasoline Prices from June 1978 in Various European Countries, 1978-80, as Compared to Increase in the Contract Price of Crude Oil and the Spot Value of Crude Oil (*cents per gallon*).

|  | Country | | | | Crude Oil | |
| --- | --- | --- | --- | --- | --- | --- |
| Month | West Germany | France | United Kingdom | Italy | Contract | Spot Value |
| November 1978 | 9.4 | NA | NA | NA | 0.0 | 6.5 |
| January 1979 | 10.9 | 2.0 | -2.0 | 2.0 | 0.2 | 13.5 |
| June 1979 | 21.8 | 16.0 | 36.0 | 2.0 | 16.9 | 56.2 |
| September 1979 | 40.3 | 22.0 | 40.0 | 20.0 | 22.5 | 47.7 |
| January 1980 | 41.8 | 50.0 | 56.0 | 52.0 | 45.6 | 57.8 |
| April 1980 | 58.1 | 49.0 | 43.0 | 49.0 | 51.7 | 43.4 |
| June 1980 | 62.1 | 62.0 | 41.0 | 66.0 | 57.6 | 49.4 |

Note: Calculations exclude changes in taxes. Data for April 1979 are not available.

Source: Product prices are taken from the CIA publication *International Energy Statistical Review* (various issues 1979-80). Prices are in current dollars measured at the rate of exchange prevailing at the time. Crude prices are taken from *Petroleum Intelligence Weekly* (Feb. 2, 1981).

Table 4-6. Cumulative Increase in Consumer Prices of Diesel Fuel from June 1978 in Various European Countries as Compared to Increase in the Contract Price of Crude Oil and the Spot Value of Crude Oil (cents per gallon).

| | Country | | | | African Light Crude | |
|---|---|---|---|---|---|---|
| Month | West Germany | France | United Kingdom | Italy | Contract | Spot Value |
| November 1978 | 9.2 | NA | NA | NA | 0.0 | 6.5 |
| January 1979 | 9.5 | 10.0 | -9.0 | 2.0 | 0.2 | 13.5 |
| June 1979 | 11.7 | 20.0 | 36.0 | 2.0 | 16.9 | 56.2 |
| September 1979 | 43.6 | 24.0 | 41.0 | 20.0 | 22.5 | 47.7 |
| January 1980 | 43.6 | 50.0 | 41.0 | 48.0 | 45.6 | 57.8 |
| April 1980 | 57.8 | 53.0 | 41.0 | 56.0 | 51.7 | 43.4 |
| June 1980 | 63.8 | 66.0 | 41.0 | 72.0 | 57.6 | 49.4 |

Note: Calculations exclude changes in taxes. Data for April 1979 are not available.

Source: Product prices are taken from the CIA publication *International Energy Statistical Review* (various issues 1978-80). Prices are in current dollars measured at the rate of exchange prevailing at the time. Crude prices are taken from *Petroleum Intelligence Weekly* (Feb. 2, 1981).

product values set in commodity markets in Rotterdam. This conclusion is reenforced by a final set of data on "inland product realizations at refinery gates" which are published from time to time by *Petroleum Intelligence Weekly* (Oct. 29, 1970; Aug. 18, 1980). These data, which are collected for the EEC, provide information on wholesale refinery prices of principal petroleum products in the major refinery centers in Europe. We have tabulated the cumulative increase in these prices between the fourth quarter of 1978 and the second quarter of 1980 (the changes are all calculated with reference to the third quarter of 1978). The data are shown on Table 4-7, which shows, for instance, that the price of naptha was $24.40 a barrel higher in 1980:2 than in 1978:3.

The pattern in cumulative increases in the price of regular gasoline, gas oil, and low sulfur residual fuel oil appears identical to that for other products as shown in the CIA data. It appears that wholesale prices follow contract crude prices and not spot market values. Naptha prices appear, however, to be more volatile.

To provide a better overall comparison of movement in prices, an average of the price increases of the four products was once again

Table 4-7. Cumulative Increases in European Refinery Prices from 1978:3 as Compared to Increases in the Cost of Crude and the Spot Value of Crude Oil, 1978:4 to 1980:2 (*dollars per barrel*).

|        |        | Products            |         |                                  | Crude Oil (African Light) |               |
| ------ | ------ | ------------------- | ------- | -------------------------------- | ------------------------- | ------------- |
| Date   | Naptha | Regular Gasoline    | Gas Oil | Low Sulfur Residual Fuel Oil     | Contract                  | Spot Value    |
| 1978:4 | 2.20   | 1.00                | 1.60    | 1.30                             | 0.00                      | 2.22          |
| 1979:1 | 6.15   | 1.60                | 4.95    | 3.70                             | 1.00                      | 9.50          |
| 1979:2 | 11.65  | 4.40                | 7.80    | 5.20                             | 5.78                      | 18.36         |
| 1979:3 | 17.15  | 11.00               | 13.35   | 8.60                             | 9.51                      | 19.28         |
| 1979:4 | 18.65  | 12.43               | 15.93   | 11.70                            | 12.24                     | 23.99         |
| 1980:1 | 26.65  | 17.65               | 22.00   | 15.33                            | 20.77                     | 21.72         |
| 1980:2 | 24.40  | 19.58               | 23.58   | 15.40                            | 22.82                     | 19.77         |

Note: Product prices represent "inland product realizations at refinery gates."

Source: Product data were published in *Petroleum Intelligence Weekly* 18, no. 44 (Oct. 29, 1979), 19, no. 33 (Aug. 18, 1980).

computed using weights that reflect the average refinery yield of these products from a barrel of African light. The cumulative increase in the weighted average is shown on Table 4-8. The second and third columns of Table 4-8 show the difference between the cumulative increase in the average product prices and the increase in the official price and the spot value of African light. The results demonstrate again that average refiner prices moved very closely with the official price of African light, with the largest difference being $2.94 a barrel in the first quarter of 1979. On the other hand, changes in average refiner prices diverged widely from movements in the spot value of African light.

To show this relationship more clearly, the movements of inland product realizations (refinery prices) in West Germany (weighted by the refinery yield for African light) are compared to the spot values of African light and the official price of African light in Figure 4-3. Once again, the close relationship between prevailing prices in product market and the official price of crude is demonstrated and the irrelevance of 1979 spot values to 1979 consumer prices is clear. This is important since, of all the western economies, West Germany's petroleum has been the least encumbered by regulation.

**Figure 4-3.** Comparison of Official Price of African Light, Spot Value of African Light, and Inland Refiner Realizations in West Germany.

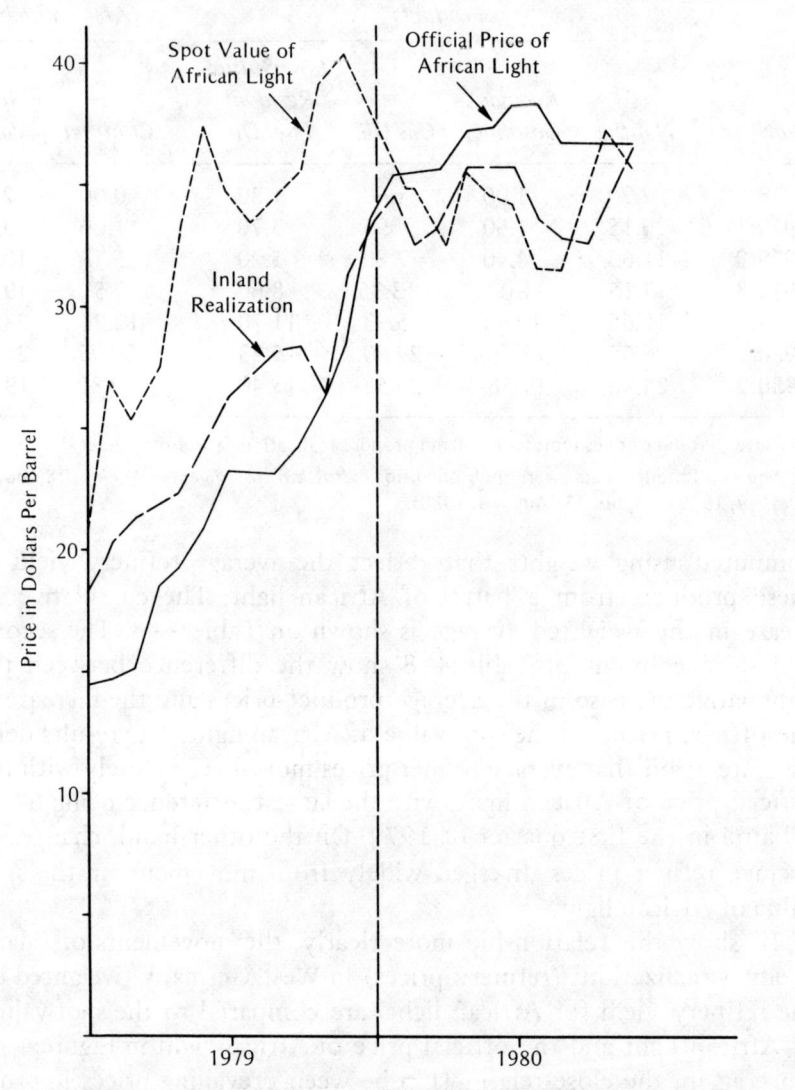

Source: Calculations by the author based on Table 4-7.

Table 4-8. Cumulative Increase from 1978:3 in Weighted Average of European Refinery Product Prices as Compared to Increases in African Light Crude Prices, 1978:4 to 1980:2 (*dollars per barrel*).

| Quarter | Cumulative Increase in Weighted Average Price | Difference Between Cumulative Increase in Weighted Average Wholesale Prices and Cumulative Increases in Crude Costs | |
|---|---|---|---|
| | | Contract Price African Light | Spot Value of African Light |
| 1978:4 | 1.40 | +1.40 | − 0.82 |
| 1979:1 | 3.93 | +2.93 | − 5.57 |
| 1979:2 | 6.55 | +0.77 | −11.81 |
| 1979:3 | 11.51 | +2.00 | − 7.77 |
| 1979:4 | 13.79 | +1.55 | −10.20 |
| 1980:1 | 19.02 | −1.75 | + 2.70 |
| 1980:2 | 19.88 | −2.94 | + 0.11 |

Note: The cumulative increase in weighted average product price represents the weighted average (using weights from a hydroskimming refinery) of product prices on Table 4-7. The difference in the first and second columns, and between the first and third columns represents the difference between the weighted average increase in product prices and the cumulative increase in crude prices shown on Table 4-7.

## CONCLUSION

This review of the history of price regulation in the United States during the 1970s and examination of the movement of consumer prices of petroleum products in the United States and the principal European countries during the two disruptions in the 1970s show that consumer prices did not change with the spot market but followed changes in official sales prices of crude oil. This conclusion may not be surprising but it is extremely important, because the conclusions reached in Chapters 3 and 4 taken together imply that any sudden change in supply will create a temporary condition which, due to the structure of the market, makes it profitable to build inventories. Specifically, a sudden increase in commodity market prices will create a profitable spread between the current and future price of crude oil. It is expected that the spread between official crude prices and spot values influences inventory behavior.

# 5 INVENTORIES

> In the aggregate, refiners drew down stocks at a moderate pace in March, April and May. However, as gasoline lines and spot shortages began to occur in May and June, refiners reduced the amount of gasoline stocks available for supply. In the aggregate, there was a net buildup of gasoline stocks during June and July. During this period, DOE repeatedly criticized refiners for overcautious stock management practices and urged more liberal use of stocks to relieve the supply situation."
>
> U.S. Department of Energy, *Final Report to the President on the Oil Supply Shortage of 1979* (1980a: 34).

No issue received more attention during the last two crises than the petroleum industry's management of inventories. During both the 1973 and 1979 crises, refiners were criticized extensively for their every action, from running down stocks too quickly to being too cautious in their inventory management. After each crisis refiners were subjected to endless abuse for compounding the problems experienced during the crisis by allowing their inventories to fall to unacceptable levels prior to the crisis. The quote from *The Economist* at the beginning of Chapter 1 sums up the criticism by accusing oil companies of selling during times of glut and acquiring inventories at times of shortage.

Nowhere in all the verbiage written about the events surrounding shortages, however, does one find an economic explanation for the inventory behavior observed during a disruption. Despite the importance of the issue, there have been no attempts to model the inventory accumulation process that occurs during a disruption and only a few analyses of the data themselves.

This chapter attempts to describe the process of inventory accumulation during disruptions in a framework consistent with the standard models of inventory accumulation. The results presented in the final section of the chapter suggest that the oil industry has followed inventory policies that represent rational responses to the economic situation. The earlier sections address the behavior that has been observed during past crises and the models of inventory behavior that have been developed by economists in their study of economic behavior in general.

## THE BEHAVIOR OF INVENTORIES DURING THE CRISES OF 1973 AND 1979

Petroleum stocks are held at five different locations between production and consumption. The first location is comprised of the point of production and the logistic system for moving the crude to the refinery. The second location is the refinery. After refining, the petroleum products become inventories held at the refinery, in transit, or at bulk plants. The fourth location where stocks are held is at retail stations and jobbers When products are sold, they move to the final location—the consumer's tank.

Statistics on petroleum inventories in the United States and other Organization for Economic Cooperation and Development (OECD) member countries provide complete information on the volume of oil held in the second and third stages and information on the volume of crude held in the first stage when it is produced in the OECD. However, no information is available on inventories held by consumers or inventories of crude in transit from non-OECD sources.

These gaps make it difficult to describe with any accuracy the changes in stocks that occur during a disruption. The usual approach to examining the behavior of stocks during disruptions has been to review the flow of oil from wells and from imports through the refinery to ultimate sales. Comparisons are made of the pattern of distribution for a specific quarter in the year of a disruption and the pattern of distribution for the previous year. As Danielson (1979: 243-63) notes, it is necessary to make a comparison on a year-over-year basis because there is a pronounced seasonal pattern to petroleum consumption.

In assessing the quarter-to-quarter movement of inventories during a crisis, it must be noted that an apparently anomalous change in stocks in one quarter may result from one of several circumstances. For instance, inventories may be increased in one quarter because a refiner anticipates a reduction in the flow of oil in the following quarter. (This would occur if the levels of inventories in transit from non-OECD sources were reduced because the loss of these stocks would not be reported.)

Precisely this type of situation occurred in 1973 and 1979 because a full six weeks are required to move oil from the Persian Gulf to storage facilities in OECD areas. Another cause of seemingly anomalous behavior in stock levels may be the weather. Warm winter weather or cold summer weather may cause consumption to fall below projected levels and cause an unanticipated increase in inventories. With these caveats noted, we turn to an evaluation of the behavior of stocks in the United States and the OECD in 1973 and 1979.

### The 1973 Crisis

Oil production was reduced by the members of the Organization of Arab Petroleum Exporting Countries (OAPEC) in early November 1973. The reduction in receipts was not noted by OECD countries until the first quarter of 1974.

In the United States deliveries of crude oil to refineries were 624,000 barrels a day (5.4 percent) higher during the fourth quarter of 1973 than during the fourth quarter of 1972, while imports of products increased by 400,000 (15.9 percent).[1] (See Table 5-1.)

---

1. Domestic crude oil production, imports of crude oil, production and imports of natural gas liquids and condensate, and the change in inventories of crude oil and intermediate products (unfinished oils and natural gasoline) were added to develop estimates of the sources of supply. Exports of crude oil then were subtracted to obtain an estimate of total inputs. The estimate of refinery output is mine. Neither the U.S. Department of Energy (DOE) nor the U.S. Bureau of Mines publishes an aggregate statistic on the output of refineries. They do, however, publish data on total deliveries, changes in product inventories, and imports and exports of products. By adding these, an estimate of refinery output was de-

Table 5-1. Sources of Refined Products Supplied to the U.S. Market and Stocks of Crude Oil and Products, 1972:4 and 1973:4.

|  |  | 1972:4 | 1973:4 |
|---|---|---|---|
| *Input and Output of Refineries* | | Thousands of Barrels a Day | |
| Domestic crude oil production | | 8,969 | 8,729 |
| Plus: | Imported crude oil | 2,496 | 3,360 |
| | Withdrawals from (additions to) refineries' inventories of crude | 48 | (13) |
| | Withdrawals from (additions to) refineries' inventories of intermediate products | 133 | 26 |
| | Natural gas liquids and condensate | 2,439 | 2,423 |
| Minus: | Exports of crude oil | (0) | (2) |
| EQUALS: | Refinery input | 14,085 | 14,523 |
| Plus: | Refinery gains and discrepancy | 466 | 439 |
| EQUALS: | Refinery output | 14,551 | 14,962 |
| *Sources of Petroleum Products Supplied to the Market* | | | |
| Refinery output | | 14,551 | 14,962 |
| Plus: | Imports of refined products | 2,523 | 2,923 |
| Minus: | Exports of refined products | 240 | 216 |
| EQUALS: | Products available to the market | 16,834 | 17,669 |
| Plus: | Withdrawals from (additions to) refineries' inventories of refined products | 767 | 67 |
| EQUALS: | Products supplied to the market | 17,601 | 17,736 |

Source: U.S. Department of Interior; Bureau of Mines; *Crude Petroleum, Petroleum Products and Natural Gas Liquids* (various issues).

rived. Because data on refinery output and inputs do not balance, a term "Refinery gains and discrepancy" was included to take into account the volumetric expansion caused by refining.

Such an increase would have permitted refiners to increase the volumes of product distributed to the market by more than 1 million barrels a day without adversely affecting their stocks (assuming they began the fourth quarter of 1973 with stocks at levels identical to those of 1972). However, refiners increased the distribution of products by only 135,000 barrels a day while using the remaining increase in supply to build stocks.

This action was justified by the embargo and was called for by the federal government. The reduction in supply did not cause domestic shortages because consumers apparently responded to appeals for conservation. For instance, *The New York Times* (Jan. 18, 1974:1) reported a 14 percent reduction in driving in New York in November 1973.

The shortage occurred in the first quarter of 1974 when the data indicate that refiner receipts of crude were reduced by 800,000 barrels a day relative to 1973:1, while product imports were reduced by 490,000 barrels a day. (See Table 5-2.) Refiners compensated for the reduced supplies by cutting supplies to the market by 1.412 million barrels a day, effectively passing the shortage on to consumers barrel for barrel. They also increased their inventories in the same way as they had during the previous year. During this time period many shortages were reported, and the first filling station lines occurred.

The distribution of products to the market did not return to pre-embargo patterns in 1974:2 despite the end of the embargo. As one can observe from Tables 5-2 and 5-3, refiners' stocks at the end of the embargo (in 1974:2) were 7 percent greater than at the end of 1973:2. Furthermore, refiner receipts of crude and product imports during 1974:2 exceeded receipts in 1973:2 by 120,000 barrels a day, yet products distributed to the market during the period decreased by 400,000 barrels a day (2.5 percent), because refiners accumulated stocks during these three quarters. This behavior was noted by many and caused considerable criticism.

The continued reduction in the distribution of products to consumers in the second quarter does not seem sensible initially, because the embargo officially ended at the end of March and many exporters had increased production before then. However, the decline may have been caused by a drop in demand as a result of emergency conservation, a slowdown in economic growth, higher prices, or a combination of all three factors. Whatever the reason, market supply and

Table 5-2. Sources of Refined Products Supplied to the U.S. Market and Stocks of Crude Oil and Products, 1973:1 and 1974:1.

|  | 1973:1 | 1974:1 |
|---|---|---|
| *Input and Output of Refineries* | \multicolumn{2}{c}{*Thousands of Barrels a Day*} | |
| Domestic crude oil production | 8,836 | 8,590 |
| Plus: Imported crude oil | 2,924 | 2,368 |
| Withdrawals from (additions to) refineries' inventories of crude | 25 | (24) |
| Withdrawals from (additions to) refineries' inventories of intermediate products | (31) | (91) |
| Natural gas liquids and condensate | 2,385 | 2,384 |
| Minus: Exports of crude oil | 0 | 9 |
| EQUALS: Refinery input | 14,139 | 13,218 |
| Plus: Refinery gains and discrepancy | 434 | 470 |
| EQUALS: Refinery output | 14,573 | 13,688 |
| *Sources of Petroleum Products Supplied to the Market* | | |
| Refinery output | 14,573 | 13,688 |
| Plus: Imports of refined products | 3,170 | 2,680 |
| Minus: Exports of refined products | 231 | 193 |
| EQUALS: Products available to the market | 17,512 | 16,175 |
| Plus: Withdrawals from (additions to) refineries' inventories of refined products | 804 | 729 |
| EQUALS: Products supplied to the market | 18,316 | 16,904 |

Source: U.S. Department of Interior, Bureau of Mines, *Crude Petroleum, Petroleum Products and Natural Gas Liquids* (various issues).

demand appear to have been in balance in the second quarter of 1974 because there are no published reports of shortages in either the general or the trade press.

It is impossible to make a similar assessment of the situation in either Japan or Europe because comparable data on petroleum consumption, imports, and stocks for Europe and Japan are not available

Table 5-3. Sources of Refined Products Supplied to the U.S. Market and Stocks of Crude Oil and Products, 1973:2 and 1974:2.

|  | 1973:2 | 1974:2 |
|---|---|---|
| *Input and Output of Refineries* | \multicolumn{2}{c}{*Thousands of Barrels a Day*} | |
| Domestic crude oil production | 8,834 | 8,481 |
| Plus: Imported crude oil | 3,162 | 3,702 |
| Withdrawals from (additions to) refineries' inventories of crude | (52) | (265) |
| Withdrawals from (additions to) refineries' inventories of intermediate products | (81) | (139) |
| Natural gas liquids and condensate | 2,409 | 2,356 |
| Minus: Exports of crude oil | 1 | 3 |
| EQUALS: Refinery input | 14,271 | 14,132 |
| Plus: Refinery gains and discrepancy | 597 | 490 |
| EQUALS: Refinery output | 14,868 | 14,622 |
| *Sources of Petroleum Products Supplied to the Market* | | |
| Refinery output | 14,868 | 14,622 |
| Plus: Imports of refined products | 2,365 | 2,313 |
| Minus: Exports of refined products | 241 | 240 |
| EQUALS: Products available to the market | 16,992 | 16,695 |
| Plus: Withdrawals from (additions to) refineries' inventories of refined products | (646) | (773) |
| EQUALS: Products supplied to the market | 16,346 | 15,922 |

Source: U.S. Department of Energy, Bureau of Mines, *Crude Petroleum, Petroleum Products and Natural Gas Liquids* (various issues).

for 1972. However, we can compare the quarter-to-quarter changes in consumption and imports in the United States, Japan, and Western Europe. These data show a pattern in Japan and Europe similar to that in the United States. (See Table 5-4.) For instance, first quarter consumption declined by 800,000 barrels a day in the United States (from 1973:4) while it declined by 1.3 million barrels a day in Eu-

**Table 5-4.** Comparison of Quarter-to-Quarter Changes in Supply and Consumption in the United States, Japan and Europe, 1973 and 1974 (*thousands of barrels a day*).

|  | United States | Europe | Japan |
|---|---|---|---|
|  | *From 1973:4 to 1974:1* | | |
| Change in consumption | −800 | −1300 | 0 |
| Change in production | −139 | 0 | 0 |
| Change in imports | −1212 | −300 | −300 |
|  | *From 1974:1 to 1974:2* | | |
| Change in consumption | −1000 | −800 | −800 |
| Change in production | −100 | 0 | 0 |
| Change in imports | 920 | 600 | 300 |

Source: CIA *International Oil Developments*.

rope and not at all in Japan. U.S. imports declined by 1.2 million barrels a day (from 1973:4), while Japanese and European imports declined by 300,000 barrels a day.[2]

From these estimates it is possible to conclude that the shortage was met primarily by reductions in consumption rather than by reductions in inventories in all three regions, especially during the first quarter.

## The 1979 Experience

The United States experienced the second petroleum crisis of the decade in 1979. American consumers were told that the cause of the crisis was a decline in Iranian oil production from 5.8 million barrels a day in July 1978 to 445,000 barrels a day in January 1979. The short-run consequences of the crisis were shortages of diesel fuel and gasoline during the months of May, June, and July. At the peak of

---

2. Note that these comparisons are from 1973:4 to 1974:1 and thus not comparable to the year-over-year comparisons on Tables 4-3, 4-5, and 4-6.

the crisis the gasoline lines were as long as or longer than those in 1974. There were also suggestions that heating oil would be in short supply in the winter of 1979-80.

The analysis here focuses on the behavior of petroleum flows. It indicates that normal levels of refinery input and petroleum production and modest price increases occurred during the first quarter of 1979. In the second quarter, however, the output of refined products fell short of projected sales. In addition, the distribution of refined products was substantially curtailed. The results were a rapid buildup of inventories and shortages.

Although the year began with inventories 7 percent below levels of a year earlier, there was no unusual buildup of inventories in the refinery sector during the first quarter of 1979. The available data summarized in Table 5-5 suggest that petroleum flows in 1979:1 closely paralleled those in 1978:1. Compared with 1978:1, refinery output in 1979:1 increased by 390,000 barrels a day or 2.4 percent, while distribution of products to the market increased by 260,000 barrels a day or 1.3 percent. Inventory changes and levels of product imports and exports were also almost identical in the two quarters. Although production of crude oil in the continental United States declined in 1979:1, total domestic crude oil production matched the 1978:1 level as Alaskan production increased from 993,000 to 1.326 million barrels a day. Exports of crude oil also increased by 221,000 barrels a day in 1979, continuing a trend that began in 1978.[3]

The supply of products during the first quarter of 1979 was apparently adequate to satisfy consumer needs; there were no reports of shortages in either the petroleum or financial press. In addition, the supply of 20.297 million barrels a day matched almost exactly the projections of required supply made by the *Oil and Gas Journal* in January 1979 and by the Independent Petroleum Association of America (IPAA) in October 1978. The IPAA projected domestic requirements during 1979:1 of 20.313 million barrels a day while the *Oil and Gas Journal* projected 20.220 million barrels a day (IPAA 1978, Lange 1979).

---

3. The increase in exports resulted from the movement of Alaskan crude oil to the U.S. Virgin Islands for refining. Sales to a refiner in the Virgin Islands are (by convention) counted as exports, and receipts from the islands as imports. The export of Alaskan oil was therefore offset by increased imports of refined products from the Virgin Islands and implies a corresponding decline in imports from other sources.

Table 5-5. Sources of Petroleum Products Supplied to the U.S. Market, 1978:1 and 1979:1.

|  | 1978:1 | 1979:1 |
|---|---|---|
|  | Thousands of Barrels a Day | |
| *Input and Output of Refineries* | | |
| Domestic crude oil production | 8,151 | 8,151 |
| Plus: Imported crude oil[a] | 5,845 | 6,333 |
| Withdrawals from (additions to) refineries' inventories of crude oil | (61) | (84) |
| Withdrawals from (additions to) refineries' inventories of intermediate products | (19) | 14 |
| Natural gas liquids and condensate | 1,983 | 2,139 |
| Minus: Exports of crude oil | 57 | 278 |
| Plus: Refinery gains and discrepancy | 397 | 352 |
| EQUALS: Refinery output | 16,238 | 16,628 |
| *Sources of Petroleum Products Supplied to the Market* | | |
| Refinery output | 16,238 | 16,628 |
| Plus: Imports of refined products | 2,188 | 2,175 |
| Minus: Exports of refined products | 189 | 216 |
| EQUALS: Products available to the market | 18,238 | 18,587 |
| Plus: Additions to refineries' inventories of refined products | 1,799 | 1,709 |
| EQUALS: Products supplied to the market | 20,037 | 20,297 |

a. Excludes imports for Strategic Petroleum Reserve.

Source: U.S. Department of Energy *Monthly Petroleum Statement*, and *Monthly Energy Review*.

The similarity between the patterns of distribution in 1978 and 1979 ended during the second quarter of 1979 when stocks were rebuilt. Indeed, the amount of inventory building was astonishing, if the preliminary data are to be believed. During the second quarter, deliveries to the market were cut 486,000 barrels a day below the

Table 5-6. Sources of Refined Products Supplied to the U.S. Market and Stocks of Crude Oil and Products, 1978:2 and 1979:2.

| | 1978:2 | 1979:2 |
|---|---|---|
| *Input and Output of Refineries* | \multicolumn{2}{c}{*Thousands of Barrels a Day*} | |
| Domestic crude oil production | 8,421 | 8,155 |
| Plus: Imported crude oil[a] | 5,668 | 6,229 |
| Withdrawals from (addition to) refineries' inventories of crude oil | 133 | (93) |
| Withdrawals from (addition to) refineries' inventories of intermediate products | 26 | (46) |
| Natural gas liquids and condensate | 1,972 | 2,082 |
| Minus: Exports of crude oil | 137 | 221 |
| Plus: Refinery gains and discrepancy | 603 | 550 |
| EQUALS: Refinery output | 16,686 | 16,656 |
| *Sources of Petroleum Products Supplied to the Market* | | |
| Refinery output | 16,686 | 16,656 |
| Plus: Imports of refined products | 1,800 | 1,655 |
| Minus: Exports of refined products | 212 | 245 |
| EQUALS: Products available to the market | 18,273 | 18,066 |
| Plus: Addition to refineries' inventories of refined products | (223) | (502) |
| EQUALS: Products supplied to the market | 18,050 | 17,564 |

a. Excludes imports for the Strategic Petroleum Reserve.
Source: Department of Energy *Monthly Petroleum Statement.*

levels of 1978:2, despite an increase in the amount of crude oil received by refineries. (See Table 5-6.)

Imports of crude oil actually increased by 561,000 barrels a day (9.9 percent) from levels a year earlier. This increase offset a decline in domestic production of 266,000 barrels a day. The total potential supply to refineries rose 321,000 barrels a day (2.0 percent), but actual throughputs rose by less than 0.5 percent because inventories of crude oil and intermediate products were increased.

Management of product inventories was also conservative. During the second quarter of 1978, refiners increased stocks of refined products at a rate of 223,000 barrels a day. During 1979, the data show that stocks were increased at a rate of 502,000 barrels a day. This difference alone reduced the supply of products available to the market by 279,000 barrels a day, or by 1.5 percent of the 1978:2 supply.

The difference in stock management between 1978:2 and 1979:2 can be summarized as follows:

|  | *Thousands of Barrels a Day* |
|---|---|
| Slower withdrawals from gasoline inventories | 337 |
| Faster inventory buildup of heating oil | 99 |
| Faster drawdown of other product inventories | -157 |
| Sum: Additional product inventories | 279 |
| Faster buildup of inventories of crude and intermediate oils | 298 |
| Total | 577 |

Thus, the differences in inventory management more than account for the reduction of 486,000 barrels a day from the 1978:2 level of supply.

Allowing for economic growth, the shortage during the second quarter was about 1 million barrels a day, according to projections by IPAA and the *Oil and Gas Journal*.[4] That the gap could have been partially closed is evident from the discussion above. If inventory changes during the second quarter had mirrored those of 1978:2, the gap would have been reduced by 577,000 barrels a day. The consequence of such a strategy would have been to begin the third quarter with total stocks of only 1,069 million barrels of crude oil and products, 86 million below the level at the same point in 1978 and 126 million below mid-1977 levels. Whether this lower level of stocks would have created an unacceptable risk of shortages later in 1979 is debatable, although DOE was worried that it would.[5] Part of their

---

4. *Oil and Gas Journal* projected domestic deliveries of 18.577 million barrels a day during the second quarter; IPAA estimated 18.5 million barrels a day. See Lange, "Industry Facing Still Larger Numbers," p. 111, and "Report of the Supply and Demand Committee."

5. In April, DOE proposed to implement programs that would have reduced U.S. consumption in the second half of 1979 below levels of a year earlier. Even without such pro-

concern may have been the uncertainty about world supplies of crude oil for the second half of the year.

The shortages experienced by the United States in 1979 do not appear as strongly in the data for Western Europe and Japan, although data published for these regions by the OECD do indicate that stock building did take place in both the first and second quarters.

In Europe (Table 5-7), total supply of crude oil and products increased by approximately 1.2 million barrels a day in the first quarter of 1979 relative to the first quarter of 1978. Of this increase, more than 60 percent (800,000 barrels a day) was distributed to final consumers, while roughly 40 percent (400,000 barrels a day) was added to stocks. Total supply during the second quarter increased by 1.2 million barrels a day relative to 1978:2. However, the distribution of the increased supply between increased stocks and increased consumption was somewhat different, with approximately 500,000 barrels a day being distributed to final demand and 900,000 barrels a day being added to inventories.

The data for Japan (Table 5-8) show an even stronger pattern. During both the first and second quarters of 1979 the available supply of crude oil and products increased relative to the same quarter of the previous year, but the increase was not adequate to meet consumer demand. In both the first and second quarters the rate of inventory accumulation therefore was lower than in the corresponding quarter of 1978. In the first quarter of 1979 inventories were reduced at a rate of approximately 110,000 barrels a day in order to meet an increase in distribution to final demand of 240,000 barrels a day. During the second quarter distribution to final sales was 230,000 barrels a day greater than during the second quarter of 1978. Forty percent (200,000 barrels a day) of this increase was met by more rapid drawdown in stocks (or slower increases). Apparently Japan, in contrast to the United States, used stocks to fill part of the shortage created by the loss in Iranian production.

---

grams, higher prices would have had an increasingly negative impact on consumption. It can be argued that a lower level of stocks would have provided the necessary buffer to support projected consumption in the third and fourth quarters, except in the event of another Iranian cutback. If such an interruption occurred, the Strategic Petroleum Reserve of 90 million barrels could have been drawn down. By September 1979, DOE was able to recover oil from the strategic reserve at a rate of 1 million barrels a day.

Table 5-7. Sources of Refined Petroleum Products Supplied to the European Market, 1978:1 versus 1979:1 and 1978:2 versus 1979:2 (*thousands of barrels a day*).

|  |  | 1978:1 | 1979:1 | Difference |
|---|---|---:|---:|---:|
| Domestic crude oil production | | 1,524 | 2,026 | 502 |
| Plus: | Imports of crude oil | 11,728 | 12,828 | 1,100 |
| Minus: | Exports of crude oil | 794 | 1,072 | 278 |
| EQUALS: | Available supply of crude oil | 12,458 | 13,782 | 1,324 |
| Minus: | Increase in stocks of crude oil | (652) | (371) | 281 |
| Plus: | Losses | 92 | (35) | (137) |
| EQUALS: | Refinery output | 13,202 | 14,118 | 916 |
| Plus: | Imports of products | 2,817 | 3,193 | 376 |
| Minus: | Exports of products | 2,175 | 2,764 | (589) |
| Minus: | Increase in stocks of products | (1,112) | (1,209) | (97) |
| EQUALS: | Distributed to final demand | 14,955 | 15,757 | 802 |

|  |  | 1978:2 | 1979:2 | Difference |
|---|---|---:|---:|---:|
| Domestic crude oil production | | 1,657 | 2,248 | 591 |
| Plus: | Imports of crude oil | 11,819 | 13,096 | 1,277 |
| Minus: | Exports of crude oil | 823 | 1,211 | 388 |
| EQUALS: | Available supply of crude oil | 12,654 | 14,133 | 1,480 |
| Minus: | Increase in stocks of crude oil | 464 | 599 | 135 |
| Plus: | Losses | 56 | 63 | 7 |
| EQUALS: | Refinery output | 12,246 | 13,597 | 1,351 |
| Plus: | Imports of products | 2,818 | 3,140 | 322 |
| Minus: | Exports of products | 2,220 | 2,793 | (573) |
| Minus: | Increase in stocks of products | (126) | 512 | (638) |
| EQUALS: | Distributed to final demand | 12,970 | 13,432 | 462 |

Note: Totals may not add due to rounding.
Source: OECD *Quarterly Oil Statistics*, 1980, No. 1. Data converted at 7.33 barrels per metric ton.

The differences in the management of inventories in the United States, Japan and Europe is shown on Table 5-9. This shows that Japan was the only country to reduce inventories during the 1979 disruption and the United States was the only country in which dis-

Table 5-8. Sources of Refined Petroleum Products Supplied to the Japanese Market, 1978:1 versus 1979:1 and 1978:2 versus 1979:2 (*thousands of barrels a day*).

|  | 1978:1 | 1979:1 | Difference |
|---|---|---|---|
| Imports of crude oil plus domestic supply of crude oil | 4,988 | 5,056 | 68 |
| Minus: Increase in stocks of crude oil | 146 | (30) | (176) |
| EQUALS: Refinery output | 4,842 | 5,086 | 244 |
| Plus: Imports of products | 463 | 515 | 52 |
| Minus: Exports of products | 22 | 9 | (13) |
| Minus: Increase in stocks of petroleum products | (366) | (300) | 66 |
| EQUALS: Distributed to final demand | 5,649 | 5,892 | 243 |

|  | 1978:2 | 1979:2 | Difference |
|---|---|---|---|
| Imports of crude oil plus domestic supply | 4,588 | 4,553 | (35) |
| Minus: Increase in stocks of crude oil | 138 | (30) | (168) |
| EQUALS: Refinery output | 4,449 | 4,583 | 133 |
| Plus: Imports of products | 469 | 537 | 68 |
| Minus: Exports of products | 11 | 8 | (3) |
| Minus: Increase in stocks of petroleum products | 119 | 87 | (32) |
| EQUALS: Distributed to final demand | 4,789 | 5,024 | 236 |

Note: Totals may not add due to rounding.
Source: OECD *Quarterly Oil Statistics*, 1980, No. 1. Data converted at 7.33 barrels per metric ton.

tribution to final consumers was reduced to achieve an increase in inventories. The United States was also the only country to exhibit signs of a shortage (Danielson 1979:243-63).[6] It must also be noted that the U.S. Department of Energy exerted strong pressure on U.S.

6. Danielson suggests that consumer countries could stabilize fluctuations in oil prices by better management of their stocks and uses 1979 data to demonstrate his point. This view is correct. However, he views the problem of inventory management in 1979 as an OECD-wide problem, when the data suggest that practices varied from region to region with the worst offender clearly being the United States.

**Table 5-9.** Comparison of Change in Available Supply and Change in Stocks Between the United States and Europe, 1979:1 and 1979:2 (*thousands of barrels a day*).

|  | 1979:1 versus 1978:1 | | |
|---|---|---|---|
|  | United States | Europe | Japan |
| Increase (decrease) in available supply | 340 | 986 | 133 |
| Minus: Increase (decrease) in rate of stock accumulation | 80 | 184 | (110) |
| EQUALS: Increase (decrease) in final distribution | 260 | 802 | 243 |
|  | 1979:2 versus 1978:2 | | |
| Increase (decrease) in available supply | 91 | 1,235 | 36 |
| Minus: Increase (decrease in rate of stock accumulation | 577 | 773 | (200) |
| EQUALS: Increase (decrease) in final distribution | (486) | 462 | 236 |

Calculations from Tables 5-5, 5-6, 5-7, and 5-8.

oil companies to increase inventories, particularly inventories of heating oil. These administration actions probably made the shortage worse. (See Verleger 1979.)

## THE REASONS FOR INVENTORY ACCUMULATION

Up to this point the fluctuations of petroleum stocks during crises have been discussed without considering the motivation for building up or drawing down stocks. One way to determine whether inventory practices in the petroleum industry have been motivated by rationality, government regulation (or threat thereof), greed, panic, or other forces is to examine the behavior of the petroleum inventories within the context of the existing economic literature on the behavior of inventories.

The standard econometric model for predicting the behavior of inventories originated with Lovell (1961), who argued that firms adjust inventories slowly over time to correct for the difference between actual and desired inventories, with the rate of adjustment determined in part by differences between expected and actual sales. The level of desired inventories (the demand for inventories) was determined by anticipated sales. Lovell (1961) noted that firms correct promptly for errors in sales expectations but are slow to change their projections of future sales.

The Lovell model of inventory accumulation is described as a stock adjustment model in which firms attempt to close the gap between the current level of inventories and the desired level of inventories, while experiencing random variation in their sales. More specifically,

$$I_t - I_{t-1} = \lambda(I_t^* - I_{t-1}) + \delta(S_t^e - S_t) \; , \tag{5.1}$$

where $I_t$ measures actual inventories at period $t$, $I_t^*$ measures the desired level of inventories, $S_t$ represents the actual level of sales, and $S^e$ the anticipated level of sales.

In the early literature, the desired level of inventories is specified as some function of expected sales,

$$I_t^* = \alpha_o + \alpha_1 S_t \; . \tag{5.2}$$

More recently, however, Lieberman (1980) and others have shown that the desired level of inventories is also a function of the cost of capital, with inventories declining as the cost of carrying them increases.

The estimation of an inventory equation using econometric methods is accomplished by substituting (5.2) for $I^*$ in equation (5.1).

$$I_t - I_{t-1} = \lambda \alpha_o + \lambda \alpha_1 S_t - \lambda I_{t-1} + \delta(S_{t-1} - S_t) \; , \tag{5.3}$$

under the assumption that $S_t^e = S_{t-1}$. The results of estimation of this specification usually imply that the rate of adjustment in inventories between actual and desired stocks is very slow. For instance, Feldstein and Auerbach (1976) find that less than 6 percent of the gap between $I$ and $I^*$ is closed in a single quarter and less than 25 percent of the gap is closed in a year. At the same time they determine that inventories correct quickly to any difference between actual and anticipated sales. (They report that 95 percent of the adjustment to any shortfall is made within the quarter.) They find these results implausible and thus offer an alternative model.

Feldstein and Auerbach argue that firms respond very quickly (fully within a quarter) to adjust actual inventories to desired (or target) levels, except for "a small effect of unanticipated sales" (Feldstein and Auerbach 1976:369), but make slow adjustments to their targets. The slow adjustment in the target is attributed to a number of institutional factors such as the availability of warehouse facilities and the planning process.

Feldstein and Auerbach also introduced a second modification to earlier models of inventory demand by separately estimating the demand functions for inventories of finished goods and intermediate goods. In their model the principal determinant of the demand for inventories of finished goods is expected sales, while the determinants of demand for inventories of intermediate goods are new orders and unfilled orders. They specify a classical stock adjustment equation for target demand for inventories of both finished and intermediate goods (referred to as "materials and goods in progress").

The Feldstein/Auerbach specification is noteworthy because the authors assume that adjustment occurs rapidly, that the demand functions differ by stage of processing, and, as Lovell (1961) notes, that costs and expectations do not affect the size of inventories. This last point might seem quite remarkable, but, as Feldstein and Auerbach argue, inventories in the economy as a whole represent only one to two days of production. Therefore they suggest that firms may be more concerned about being caught short than with the cost of carrying inventories.

Two critical features of the Feldstein/Auerbach model are captured in an adjustment equation, which is specified as

$$I_t^* - I_{t-1}^* = \mu(\gamma_1 + \gamma_2 S_{t,t}^e - I_{t-1}^*) + \epsilon_t \tag{5.4}$$

where $S_{t,t}^e$ represents expected sales in period $t$ at the start of period $t$. Actual inventories are derived from an equation

$$I_t = I_t^* + \gamma_o (S_{t,t}^e - S_t) + u_t \ . \tag{5.5}$$

By substituting equation (5.5) into (5.4) and solving a lagged form of equation (5.4) for a value of $I^*$, Feldstein and Auerbach propose a model in which the current level of inventories is determined by last period's inventories, the difference between actual and realized sales last period, expected sales in the current period, and the difference between actual and expected sales in the current period. Their empirical tests of this hypothesis justify (to them) their conclusions that inventories adjust quickly to changes in expectations but slowly to changes in sales.

There have been few attempts to apply models of inventory behavior developed for the economy as a whole to the petroleum industry. One study by Griffin (1971) found that for the period from 1948 to 1966 the speed of adjustment in product inventories was approximately 10 percent per month in a specification that followed the approach used by Lovell.

The straightforward application of the standard economic theory of inventory accumulation to the petroleum industry is, however, probably inappropriate for several reasons. First, inventories play a substantially different role in oil than in other sectors of the economy; second, inventories of petroleum products are subject to much greater changes in value than inventories in other sectors of the economy; third, some portion of oil inventories are made up of pipeline "fill" and are not usable; and finally, the cost of holding oil inventories is probably greater than the cost of holding inventories in manufacturing.

The difference between the roles played by inventories in the economy as a whole and in oil may be observed by comparing the fluctuations in the level of stocks over a business cycle. With respect to inventories as a whole, Feldstein and Auerbach (1976:356) note, "Major changes in inventories represent the outputs and inputs of only very short time periods." They point out that the largest one-year increase in finished goods inventories in the United States (from 1966:2 to 1967:2) amounted to $2 billion and calculate that this increase was equal to less than two days' production. They also note that the largest change in unfinished goods inventories (from 1974:4 to 1976:1) was $4.5 billion and calculate that this was equivalent to a change of output of less than five days.

In contrast, fluctuations in inventories of petroleum products are very large. During the period from 1977 to 1980, the fluctuations in inventories has amounted to as much as 20 days during four consecutive quarters, and from the trough in 1979:1 to the peak in 1980:3 inventories increased from 70.8 days of consumption to 115 days of

Table 5-10. Days of Supply of Oil Held by OECD Countries.

| Date | Inventories of Crude | Inventories of Product | Consumption | Calculated Days of Supply |
|---|---|---|---|---|
| | Million Metric Tons | | | Days |
| 1978:1 | 156.4 | 216.8 | 457.7 | 83.4 |
| 1978:2 | 167.8 | 216.6 | 400.9 | 86.1 |
| 1978:3 | 163.6 | 238.0 | 402.4 | 90.1 |
| 1978:4 | 166.2 | 238.0 | 453.1 | 80.3 |
| 1979:1 | 161.7 | 206.6 | 468.2 | 70.8 |
| 1979:2 | 173.6 | 219.2 | 402.9 | 87.3 |
| 1979:3 | 179.2 | 254.3 | 400.0 | 97.5 |
| 1979:4 | 184.2 | 259.6 | 444.2 | 90.0 |
| 1980:1 | 184.0 | 253.6 | 439.2 | 90.0 |
| 1980:2 | 200.5 | 263.5 | 382.2 | 109.3 |
| 1980:3 | 207.2 | 278.6 | 378.9 | 115.4 |

Source: OECD *Quarterly Oil Statistics* 1979, no. 1; 1980, no. 4.

consumption. (See Table 5-10.) Thus, changes in oil inventories have a fundamentally different character than changes in inventories in other sectors of the economy.

A second characteristic of oil inventories relates to the change in the value of stocks. Prices of crude petroleum and petroleum products are volatile, while prices of other manufactured goods, both in finished form and the raw material inputs to the manufacturing process, are not. This difference can be seen by comparing the rate of increase in the price of imported crude oil with the rate of increase in the deflator for manufacturers' inventories published as part of the U.S. National Income and Product Accounts by the Department of Commerce. These two measures of inventory valuation are shown on Table 5-11. It may be noted that during the period of oil price increases in 1979 the value of oil inventories increased at rates which were eighteen times the rate of increase in manufactured inventories![7]

[7]. The price of imported crude oil is used as the measure of increase in inventory values because most refiners use a LIFO (Last in First Out) accounting procedure. The price of imported crude then provides a measure of the rate of change of the inventories of crude held by the refiners.

**Table 5-11.** Indicators of the Increase in U.S. Value of Inventories in Manufacturing and Petroleum, Quarterly from 1978 to 1980.

| Quarter | Increase in Value of Oil Inventories (Measured by Increase in Landed Cost of Imported Oil) | Increase in Value of Manufactured Inventories (Measured by Change in Price Deflator for Manufacturing Inventories) |
|---|---|---|
| | Quarterly Rate of Change at Compound Annual Rates (Change from previous quarter) | |
| 1978:2 | 0.2% | 6.7% |
| 1978:3 | -5.5 | 8.6 |
| 1978:4 | 5.9 | 11.2 |
| 1979:1 | 34.3 | 12.8 |
| 1979:2 | 112.6 | 16.1 |
| 1979:3 | 290.2 | 16.3 |
| 1979:4 | 58.9 | 20.8 |
| 1980:1 | 102.2 | 14.4 |
| 1980:2 | 27.3 | 8.2 |
| 1980:3 | 4.2 | 8.8 |

Source: Value of manufactured inventories: U.S. national income accounts deflator for inventories calculated from Tables B15 and B16 of Council of Economic Advisers, *The Economic Report of the President*, 1981. Price of Crude Oil: DOE *Monthly Energy Review* (various issues).

The potential for increases in oil prices creates an obvious incentive for firms to hold larger inventories. Opposed to this incentive, however, is the consideration of cost, including the potential for price decreases. As interest rates increase and as the cost of acquiring oil increases, firms can be expected to plan to draw down stocks. On the other hand, during times of turmoil on world oil markets (i.e., during periods when commodity prices are increasing), refiners can be expected to attempt to increase stocks if the expected increase in price exceeds the cost of holding inventories.

This incentive may be described as follows. Inventories tend to increase when the expected increase in price $E(\Delta P^e)$ between two periods is greater than the opportunity cost of carrying the oil through to the next period. This cost can be found by multiplying the interest rate by the cost of crude ($P$). Therefore, a refiner should acquire crude oil when

$$E(\Delta P^e) = P^c_{t+1} - P^c_t \geq r_t P^c_t \ . \tag{5.6}$$

Here we can apply the results of Chapter 3 by noting that the expected value of $P^c_{t+1}$ is known in advance because

$$E(P^c_{t+1}) = \alpha P^c_t + (1 - \alpha) SV_t \ . \tag{5.7}$$

Thus, a profit-maximizing firm should be expected to increase its inventories if

$$(1 - \alpha)(SV_t - P^c_t) \geq r_t P^c_t \tag{5.8}$$

or

$$(SV_t - P^c_t) \geq \frac{r_t P^c_t}{1 - \alpha} \ . \tag{5.9}$$

The result shown in equation (5.9) assumes that the physical cost of holding inventories is zero, which will not usually be the case. To correct for this omission, we add an additional element, $K$, to equation (5.6) to represent the cost of holding stocks. ($K$ is the cost in dollars of holding a barrel of oil in inventory for an additional quarter.) After making this adjustment, the cost of holding inventories becomes

$$SV_t - P^c_t \geq \frac{r_t P^c_t + K}{1 - \alpha} \ . \tag{5.10}$$

Broadly speaking, equation (5.10) suggests that, other things being equal, firms will tend to liquidate stocks of oil during periods of stable prices because the spread between spot values and the official price of crude is less than the opportunity cost of carrying a barrel of oil. In addition, they will tend to increase their inventories at times when the difference between the spot value of crude oil and the price of crude oil exceeds the cost of carrying the oil plus the physical storage cost, with the latter term adjusted for the expected rate of increase in oil prices.

The importance of $\alpha$ in these calculations should be emphasized. During those periods when spot values of crude increase, there is an incentive to increase inventories. This incentive, however, is tempered by the facts that exporters of crude oil increase their prices slowly, and the prices of the products are increased only as crude prices are increased (the point made in Chapter 4). Thus, the incentive to acquire stocks is, to an extent, dampened unless the spread between spot and contract prices is very large.

## APPLICATION OF THE THEORY OF INVENTORY ACCUMULATION TO OIL

The materials presented above suggest that we should observe a causal relationship between oil stocks, prices of crude oil, spot values of crude oil, and interest rates. This section examines the hypothesis that simple models of inventory behavior can be used to explain the acquisition and disposal of petroleum inventories. The discussion will be presented in three sections. In the first two, movements of actual data are compared to changes in the difference between spot values and the price of crude. Then we offer econometric evidence to support this conclusion.

### The Relationships Between Stocks and Profits

The relationship between inventories and market conditions predicted in this theoretical analysis has been observed in U.S. markets duing the last four years. The movement of stocks has corresponded to movements in current or expected profit from inventories. To show this relationship costs of holding inventories are computed by taking the difference between the refiner acquisition cost of crude oil (after entitlements) and the realizations available on the Gulf Coast market. (The latter represents the value of the products derived from an incremental barrel of crude oil sold on the Gulf Coast.) Both refiner acquisition costs of crude and spot values are shown on Figure 5-1. The current difference between crude costs and realizations is used as an indicator of the profit that a refiner may expect to realize by holding inventories.

According to this model of oil supply and demand, refiners would be expected to increase inventories when the difference between product realizations and crude costs increases, and decrease them when the difference between realizations and crude costs decreases (other things remaining equal). This is precisely what appears to have occurred between 1976 and 1980. For example, if the difference be-

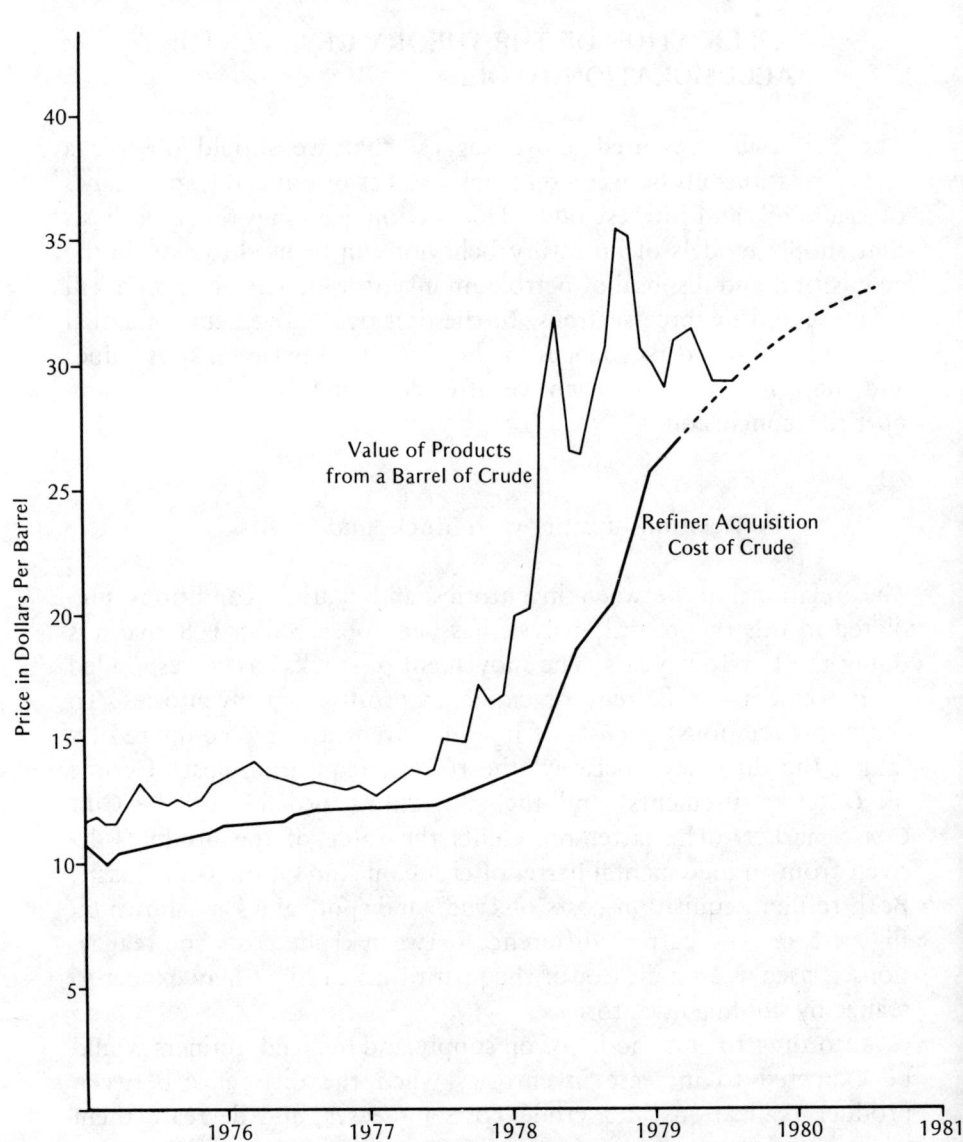

**Figure 5-1.** Comparison of Spot Market Value of Products from a Barrel of Crude on Gulf Coast Markets with the Postentitlement Refiner Acquisition Cost of Crude.

**Figure 5-2.** Comparison of Petroleum Inventories (Not Seasonally Adjusted) with the Profit on Products from a Barrel of Oil Sold at Platt's Gulf Coast Spot Prices.

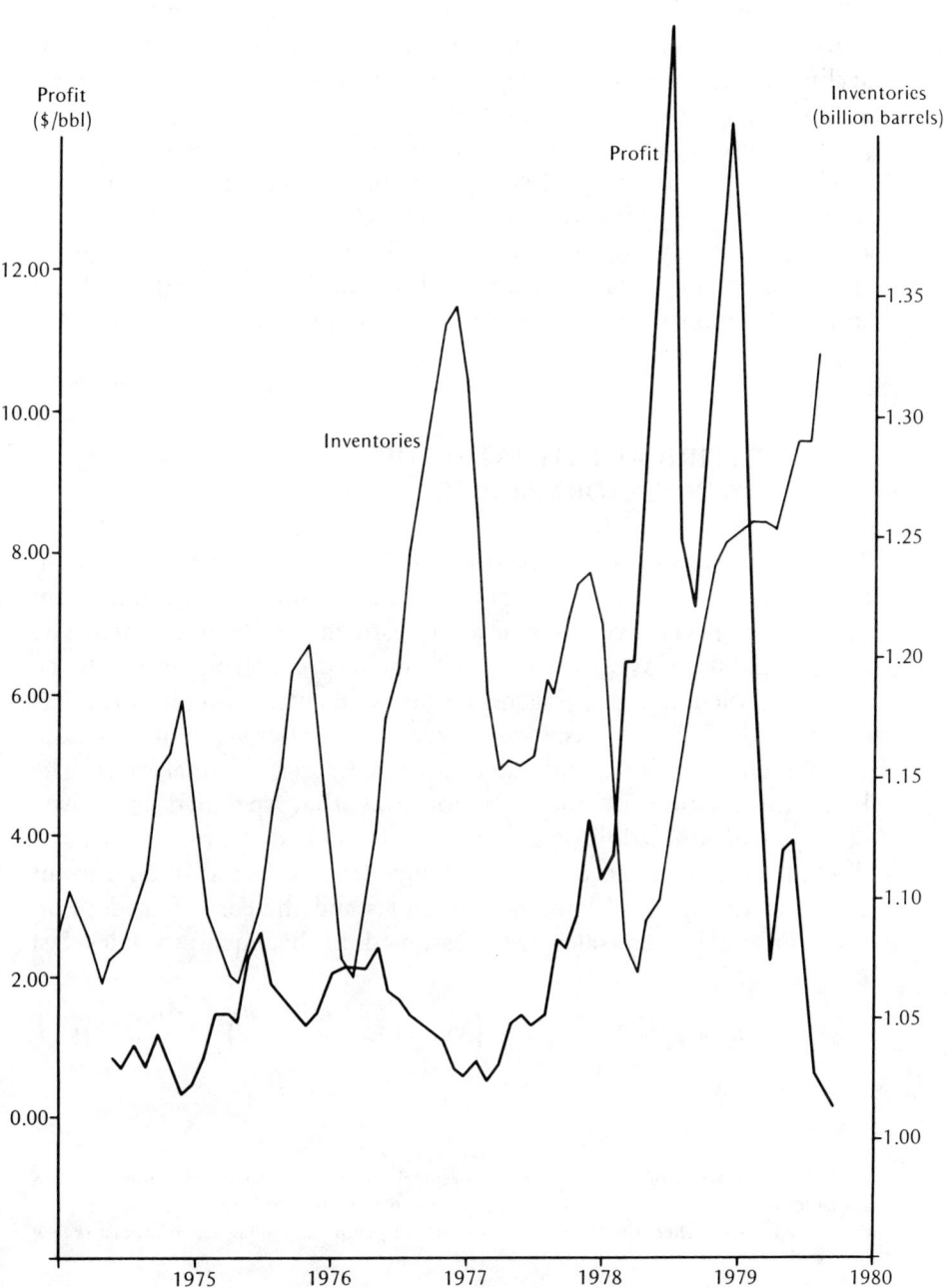

tween realizations and crude costs "gross realization" (for lack of a better name) is compared with the movement of inventories (see Figure 5-2), it can be observed that the periods of low profit in late 1977 led to a decline in inventories in early 1978. The decline in profits in the summer of 1980 appears to have set the stage for a decline in inventories that may have been frustrated by the outbreak of hostilities between Iran and Iraq.

There also appears to be a correspondence between increases in stocks and increases in profits. The comparison shown on Figure 5-2 is exceedingly rough, however, for two reasons: First, the stock data are not seasonally adjusted, and second, refiners may not have been able to move inventories to desired levels in the last months of 1978 and early months of 1979 when crude production in Iran first sagged.

## EMPIRICAL ESTIMATION OF AN INVENTORY MODEL

The second step in the analysis of the behavior of petroleum inventories was to estimate econometric demand functions for petroleum inventories. Several versions of an adjustment model were tried. The basic approach was to estimate the model for petroleum products, crude petroleum, and the aggregate of crude petroleum plus products as a function of the price spread between the spot value of crude oil, official OPEC prices, and alternative measures of final sales. The basic formulation of the equation was that specified by Lovell (1961), with one additional variable—the spread in prices. Thus, the desired level of inventories was assumed to be based upon current sales and the spread between spot values and the cost of crude. Furthermore, expected sales were assumed to be equal to sales last period ($S_t^e = S_{t-1}$):[8]

$$I_t^* = \beta_o + \beta_1 S_{t-1} + \beta_2 \left( SV_t - P_t^c - \frac{r_t P_t^c + k}{1 - \alpha} \right) . \quad (5.11)$$

---

8. In the analysis of the demand for inventories in the economy as a whole, data are available for expected sales. Similar data are not available for petroleum products, except in the United States, where the *Oil and Gas Journal* has historically published a forecast review issue each January.

The adjustment process was assumed to be given by the process

$$I_t - I_{t-1} = \rho (I_t^* - I_{t-1}) + \epsilon_t \ , \tag{5.12}$$

where the term $\epsilon$ is a random error. The estimating equation (equation (5.13)) was obtained by substituting for $I_t^*$ in equation (5.12):

$$I_t = \beta_0 \rho + \beta_1 \rho S_{t-1} + \beta_2 \rho \left( SV_t - P_t^c - \frac{r_t P_t^c}{1-\alpha} \right) + (1-\rho) I_{t-1} + \epsilon_t \ . \tag{5.13}$$

Equation (5.13) was estimated using ordinary least squares and instrumental variables estimating techniques. The equations were fitted to OECD data on inventories for the United States, Japan, Europe, and the total of the OECD. For each region three inventory equations were estimated: one for crude and products as an aggregate, one for crude oil alone, and one for product alone. In the process of estimation, no attempt was made to correct for government stockpiles.

The statistical results from the estimation of equation (5.13) were adequate but not good. The lagged sales variable was generally not statistically significant and frequently had the wrong sign. Furthermore, seasonal dummy variables were generally not statistically significant. On the other hand, the variable measuring the spread between the spot value of crude oil and the official price always entered the equations with the correct sign and was usually statistically at the 90 percent confidence level when a one-tailed $t$ test was applied.

The empirical results also caused us to reject the hypothesis that crude and product inventories should be aggregated. Instead, the model provided an adequate explanation for the process by which crude oil inventories and product inventories are accumulated, but not the aggregate of crude and product. The results of the estimation process are shown on Tables 5-12 and 5-13.

The failure of consumption to prove statistically significant in the estimation of the inventory equation was surprising. However, the period of estimation spanned the period 1974:3 to 1980:2, a period of extreme fluctuations in consumption due to substantial changes in prices, government pressures to substitute other fuels for oil, and extensive efforts to conserve. The vicissitudes of the weather also caused substantial unexpected swings in demand. These pressures caused many forecasts of oil demand to be substantially in error. It should not be surprising that sales of petroleum failed to explain the behavior of inventories.

Table 5-12. Inventory Demand Equations Estimated by Ordinary Least Squares.

| Region | Constant | Price Spread | ρ | $R^2$ | DW | SEE | Barrels Per Day Acquired or Sold (per $ spread) |
|---|---|---|---|---|---|---|---|
| *Total Inventories* | | | | | | | |
| OECD | 118114[b] (47773) | 1658[a] (1140) | .686[b] (.116) | .686 | 1.928 | 23331 | 147 |
| United States | 65436[b] (22371) | 586 (418) | .568[b] (.151) | .471 | 1.903 | 8770 | 52 |
| Europe | 43081[b] (17477) | 810 (673) | .713[b] (.122) | .661 | 2.054 | 13857 | 72 |
| Japan | 13337[b] (4412) | 243 (164) | .733[b] (.094) | .792 | 2.012 | 3296 | 21 |
| *Inventories of Crude Oil* | | | | | | | |
| OECD | 35344[b] (12252) | 724[a] (373) | .773[b] (.085) | .868 | 2.869 | 6933 | 64 |
| United States | 20301[b] (8750) | 389[a] (209) | .657[b] (.152) | .708 | 1.729 | 3500 | 35 |
| Europe | 14959[b] (6154) | 326 (224) | .712[b] (.124) | .688 | 2.697 | 4498 | 29 |
| Japan | 6256[b] (2013) | 121 (95) | .828[b] (.064) | .910 | 2.247 | 1874 | 11 |

*Inventories of Product*

|  |  |  |  |  |  |  |  |
|---|---|---|---|---|---|---|---|
| OECD | 94126[b] (31747) | 990 (854) | .575[b] (.147) | .436 | 1.760 | 18446 | 88 |
| United States | 49767[b] (15506) | 122 (327) | .461[b] (.170) | .194 | 1.815 | 7318 | 11 |
| Europe | 32541[b] (12630) | 549 (510) | .665[b] (.135) | .574 | 2.000 | 10671 | 49 |
| Japan | 12750[b] (3099) | 91 (91) | .179[b] (.198) | .004 | 1.845 | 2011 | 8 |

a. Statistically significant at a 95 percent level under a one tailed $t$ test.
b. Statistically significant at a 99 percent level.

Table 5-13. Inventory Demand Equations Estimated by Instrumental Variables.

| Region | Constant | Price Spread | $\rho$ | $R^2$ | DW | SEE | Barrels Per Day Acquired or Sold (per $ spread) |
|---|---|---|---|---|---|---|---|
| | | | | Total Inventories | | | |
| OECD | 212333[b] (46215) | 1693[a] (1160) | .677[b] (.129) | .683 | 1.913 | 23335 | 150 |
| United States | 54400[b] (21650) | 515 (429) | .643[b] (.176) | .464 | 2.000 | 8820 | 46 |
| Europe | 38982[a] (20549) | 749 (694) | .742[b] (.145) | .665 | 2.103 | 13875 | 66 |
| Japan | 16183[b] (5048) | 290 (170) | .671[b] (.108) | .788 | 1.849 | 3329 | 26 |
| | | | | Inventories of Crude Oil | | | |
| OECD | 34496[b] (13752) | 710[b] (387) | .779[b] (.095) | .868 | 2.887 | 6934 | 63 |
| United States | 13798 (12156) | 286 (249) | .771[b] (.212) | .701 | 1.840 | 3545 | 25 |
| Europe | 11144[a] (7475) | 263 (237) | .791[b] (.152) | .682 | 2.872 | 4540 | 23 |
| Japan | 7074[b] (2363) | 139[a] (99) | .801 (.760) | .909 | 2.153 | 1881 | 12 |

*Inventories of Product*

|  |  |  |  |  |  |  |
|---|---|---|---|---|---|---|
| OECD | 97873[b] | 916 | .558[b] | .436 | 1.737 | 18452 | 81 |
|  | (34997) | (861) | (.162) |  |  |  |  |
| United States | 45450[b] | 120 | .509[b] | .194 | 1.879 | 7332 | 11 |
|  | (17798) | (328) | (.195) |  |  |  |  |
| Europe | 30977[b] | 526 | .682[b] | .573 | 2.029 | 10676 | 47 |
|  | (14759) | (523) | (.159) |  |  |  |  |
| Japan | 13309[b] | 93 | .143 | .003 | 1.818 | 2013 | 8 |
|  | (3362) | (91) | (.215) |  |  |  |  |

a. Statistically significant at a 95 percent level under a one tailed *t* test.
b. Statistically significant at a 99 percent level.

The lack of statistical significance on some of the price variables is more troubling. The $t$ statistics on most are very low. In general, however, all of the coefficients are statistically significant when a one-tailed $t$ test is applied (the hypothesis being that the parameter is positive and not just different from zero). This test is accepted here because a negative sign for the parameter would be contrary to the theory in the model. Furthermore, all of the twelve parameters on the price variable are positive, which lends strength to the conclusion that the price spread is a determinant of inventories in the petroleum sector. (If the parameter were truly zero, one would expect to find that half of the statistically estimated parameters were positive and half were negative. The probability of drawing twelve of twelve positive parameters from twelve independent data sets is negligible.)

An examination of the residuals from the crude oil inventory demand equations (Tables 5-14 and 5-15) also indicates that the predicted values follow movements in actual inventories with a fairly regular pattern. In fact, the equations do very well during the 1979 embroglio.

The econometric model of fluctuations in petroleum inventories, therefore, explains some, but by no means all, of the variance in the data. This is probably all one can expect because there are a number of external forces influencing movements in oil inventories, including:

- Government policies that require firms to acquire or hold specific levels of stocks, such as the programs in France and West Germany (Deese and Miller, 1981);
- Government programs that mandate sales from large firms to small firms, such as the U.S. crude oil buy/sell program;
- Unanticipated changes in consumer demand that cause unexpected increases or decreases in consumption;
- Abnormal weather, such as warm winters or cold summers, that reduces or increases normal demands; and
- The six-week shipping lag between loading ports and refineries, which means that a large volume of oil not counted in inventories is inexorably on its way to the tanks of refiners and thus will appear in next quarter's inventories.

These and other factors explain why it is difficult to model the behavior of oil industry inventories. Nevertheless, it can be concluded

Table 5-14. Comparison of Actual and Predicted Rates of Acquisition of Stocks of Crude Oil, 1979 and 1980 (*thousand barrels a day*).

| | Region | | | |
|---|---|---|---|---|
| Quarter | All OECD | United States | Japan | Europe |
| 1979:1 | 152 | 47 | 61 | 82 |
| 1979:2 | -727 | -112 | -97 | -527 |
| 1979:3 | 293 | 83 | -80 | 276 |
| 1979:4 | 50 | 19 | -55 | 104 |
| 1980:1 | 42 | -51 | 185 | -85 |
| 1980:2 | 195 | 154 | -46 | 84 |

Note: A negative entry indicates that the predicted level of inventory acquisition was greater than the actual rate of acquisition.

Table 5-15. Comparison of Actual and Predicted Rates of Acquisition of Stocks of Products, 1979 and 1980 (*thousand barrels a day*).

| Quarter | All OECD | United States | Japan | Europe |
|---|---|---|---|---|
| 1979:1 | 735 | 333 | 79 | 162 |
| 1979:2 | -3193 | -1317 | -1447 | -296 |
| 1979:3 | -655 | -110 | -270 | -187 |
| 1979:4 | 1985 | 429 | 1213 | -255 |
| 1980:1 | 674 | 482 | 25 | 154 |
| 1980:2 | 492 | 314 | 41 | 11 |

Note: A negative entry indicates that the predicted level of inventory acquisition was greater than the actual rate of acquisition.

that the difference between spot values and contract prices does explain some of the variance in inventories.

## CONCLUSION

This chapter began by describing the behavior of inventory acquisition during the 1973-74 and 1979 disruptions, noting that inventories apparently were acquired at perverse times. That discussion seemed to support the argument of *The Economist* (September 27, 1980:16) that "Oil companies are expert at accumulating their stocks in the most disruptive way." Then, in the second half of the chapter, it was shown that there clearly were circumstances under which it would be profitable to acquire stocks. The literature on inventory modeling was applied to the question of oil industry stock acquisition, demonstrating that an econometric model of inventory behavior explained part of the process of stock acquisition. Finally, that model was used to reassess inventory acquisition in the 1979 crisis.

The means of assessment is an examination of the residuals from the inventory equations for the pertinent quarters in 1979. If these residuals are small, one would conclude that inventory management practices had not changed during the crisis in 1979. Alternatively, if the residuals are large one would conclude that the response to the crisis was to either reduce or increase stockbuilding during the crisis. As it turns out, the residuals are very large and negative for each region (OECD as a whole, Europe, Japan, and the United States) for both crude oil and products. The negative residuals imply in this case that those holding inventories (oil companies) did not achieve the econometrically projected target, leading to the conclusion that actions by the industry tended to reduce the magnitude of the disruption. Such a statement might be qualified by arguing that storage capacity was not available or that the oil was not available, but such qualifications would not be accurate.

According to this simple analysis, the stock building during the 1979 crisis was less than would be predicted by our simple econometric model. The statistics support the beleaguered oil executives, who, when accused of liquidating stocks during times of plenty and building stocks during times of shortage, claim that they were acting in the public interest. Accusations such as those by DOE (1980a:34) which blame the oil industry for being "overcautious in stock management," are simply wrong. Indeed, if any accusation is deserved, it is that oil company officials acted more in the public interest and less in the interest of their stockholders by not acquiring stocks at normal rates.

# 6 MODELING THE BEHAVIOR OF THE MARKET DURING DISRUPTIONS

> The periodic upward ratcheting of crude oil and product prices is an inherent part of the present world oil market structure. Unstable production levels by (and within) OPEC are also inherent given the present structure.
>
> Albert L. Danielsen, "The Role of Speculation in the Oil Price Ratchet Process," *Resources and Energy* 2 (1979):257.

The previous chapters have presented three-quarters of a model of the "upward ratcheting process of crude oil and product prices" described by Danielson. Chapter 2 showed that the increase in prices on commodity markets is determined by reductions in output in producing countries. Chapters 3 and 4 then showed that OPEC prices tend to follow spot values computed from commodity markets and that consumer prices have increased only with increases in the world cost of crude (OPEC prices). Finally, it was noted in Chapter 5 that demand for stocks is influenced by the spread between spot values and the official price of crude oil. This chapter ties these four distinct characteristics of the oil market together into a single model so that the process of price behavior during disruptions can be analyzed.

The simple four equation model is described in Table 6-1. Equation (6.1) specifies the relationship between current official crude

**Table 6-1. Summary of Model.**

| | | |
|---|---|---|
| $P_t^c$ | = | Official crude price in period $t$ |
| $SV_t$ | = | Spot value in period $t$ |
| $P_t^*$ | = | Consumer product price in period $t$ |
| $I_t$ | = | Demand for inventories in period $t$ |
| $Q_t^e$ | = | Expected production of oil in period $t$ |
| $Q_t$ | = | Actual production of oil in period $t$ |
| $Q_t^D$ | = | Demand in period $t$ |

Official Price of Crude Oil

$$P_t^c = \alpha P_{t-1}^c + (1-\alpha) SV_{t-1} \tag{6.1}$$

Consumer Prices for Petroleum Products

$$P_t^* = A + P_t^c \tag{6.2}$$

Demand for Inventories

$$I_t = \beta_0 + \beta_1 \left( SV_{t-1} - P_{t-1}^c - \frac{r_t P_t^c}{1-\alpha} \right) + \beta_2 I_{t-1} \tag{6.3}$$

Spot Values for Oil

$$SV_t = SV_{t-1} (Q_t^e/Q_t)^\gamma \tag{6.4}$$

*Auxiliary Equations*

Consumer Demand for Petroleum

$$\text{Log}(Q_t^D) = \delta_0 + \delta_1 \text{Log}(P_t^*) + \delta_2 \text{Log}(Y_t) + \delta_3 \text{Log}(Q_{t-1}^D) \tag{6.A1}$$

Putative Shortage/Glut Equation

$$\text{Shortage} = Q_t - Q_t^D - (I_t - I_{t-1}) \tag{6.A2}$$

Expectation Equation

$$Q_t^e = \xi(Q_t^D) + \xi(I_t - I_{t-1}) \tag{6.A3}$$

prices as a function of official crude prices in the prior period and spot values in the prior period. Equation (6.2) shows the relationship between consumer prices and official prices. Equation (6.3) is the inventory demand equation described in Chapter 4. Finally, equation (6.4) describes the relationship between current spot values, expected crude oil production, and actual crude oil production.

Three other equations are also shown in Table 6-1. The first, equation (6.A1), specifies the relationship between consumer demand for oil, economic conditions, and consumer prices. The second, equation (6.A2) is an identity that specifies the size of any shortage that may occur during a disruption. The putative shortage is defined as the difference between actual supply, consumer demand, and the change in stocks. This shortage may be compared to the reduction in the product supplied to the market, discussed in Chapter 4. Finally, equation (6.A3) indicates a general relationship between expected supply and consumer demand.

Equations (6.A1), (6.A2), and (6.A3) are introduced to complete the model; none of the three is estimated here. (Indeed, (6.A2) and (6.A3) are not estimable.) Furthermore, none is absolutely essential to the model described in equations (6.1) through (6.4). However, the three provide a means by which to analyze the effect of alternative policies designed to address disruptions.

## CONSUMER DEMAND, SHORTAGES AND CHANGES IN EXPECTED SUPPLY

Equation (6.A1) describes consumer demand for petroleum products. It is specified as an adjustment equation, where demand is written as a function of the consumer price of petroleum, $P^*$, and a measure of income, $Y$.

$$\text{Log}(Q_t^D) = \delta_0 + \delta_1 \text{Log}(P_t^*) + \delta_2 \text{Log}(Y_t) + \delta_3 \text{Log}(Q_{t-1}^D) \quad (6.A1)$$

Rather than estimate equation (6.A1), we summarize in Table 6-2 the results of income and price elasticities of demand reported by a number of other authors. These elasticities are for short- and not long-run demand and generally exclude the substitution effects one finds in other studies such as Griffin (1971) or Pindyck (1978). The exclusion of the results of these studies was intentional because they model long-run, not short-run, behavior.

**Table 6-2.** Published Estimates of Short- and Long-Run Price Elasticities of Demand for Petroleum Products.

| | | Price Elasticity Estimates | |
|---|---|---|---|
| Sector | Study | Short-Run | Long-Run |
| Residential | Phillips (1972) | −.12 | −.73 |
| Residential | Verleger and Sheehan (1974) | −.22 | −.93 |
| Residential | FEA (1976) | −.41 | −.87 |
| Commercial | FEA (1976) | — | −.56 |
| Commercial | Baughman and Joskow (1975) | −.18 | −1.12 |
| Industrial | Verleger and Sheehan (1974) | −.12 | −.61 |
| Industrial | Baughman and Zerhoot (1975) | −.11 | −1.32 |
| Industrial | FEA (1976) | | |
| | Distillate | −.34 | −1.01 |
| | Residual Fuel Oil | −.26 | −.75 |
| Transportation | Ramsey, Rasche and Allen (1975) | — | −.77 |
| Transportation | Houthakker, Verleger and Sheehan (1974) | −.08 | −.25 |
| Transportation | Verleger (1975) | −.16 | −.54 |

Equation (6.A2) is an identity that describes the imputation of a putative shortage or glut of oil,

$$\text{Shortage} = Q_t - Q_t^D - (I_t - I_{t-1}) ,  \quad (6.\text{A2})$$

where $Q_t$ represents world production of oil, $Q_t^D$ represents consumer demand (from equation (6.A1)), and $(I_t - I_{t-1})$ represents the change in stocks.

Equation (6.A3) describes the manner in which expected or anticipated production is calculated absent government regulation:

$$Q_t^e = \xi(Q_t^D) + \xi(I_t - I_{t-1}) \quad (6.\text{A3})$$

Here it is assumed that purchasers of crude oil (oil companies) make their nominations for crude oil liftings in any quarter (or month) based upon their expectations of consumer demand in the current or upcoming period plus their expected change in stocks. This equation ties back into the calculation of the behavior of spot market prices described in Chapter 2.[1]

## THE BEHAVIOR OF THE OIL MARKET

This simple model can now be used to describe the behavior of the oil market both in periods of stability and periods of disruption. During periods of tranquility, stability is expected to be observed in spot crude values. Stability in spot crude values then implies that the spot value of crude oil will be approximately equal to the official price of crude oil. The equilibrium between spot crude values and official crude prices will imply that there is no speculative demand for inventories. In addition, under these conditions we assume that expected production as given by equation (6.A3) is approximately equal to production. So long as this happens, there is no change in spot values, and the process will be repeated in the next period.

Of course it is possible that unusual weather or some other shock may cause consumption to exceed or fall short of the projected level. Alternatively, it is possible for a portion of production to be disrupted temporarily due to a mechanical problem. In this case a mini-shortage or glut could be generated from the shortage identity (6.A3). In the normal course of business such a shortage would be expected to be filled by drawing down inventories so that the actual change in inventories would either fall short of or exceed the intended adjustment. Offsetting adjustments would then be made to the projected inventory demand for the following period. As long as periodic adjustments can be made in $Q_t$ and these adjustments are reflected in $Q_t^e$, relative stability would be expected to be observed in the market.

Offsetting adjustments in supply are assumed to be impossible during a disruption. Instead, a loss in supply will create a gap between $Q_t^e$ and $Q_t$ due to a decline in actual production. During the

---

1. Note that the measurement of $Q_t^e$ here is different than in Chapter 5. Please refer to footnote 5 in Chapter 2 for an explanation of the difference.

first period this shortfall will have only one effect—to cause an increase in spot values.[2] The increase in spot values in period $t$ then starts in motion a series of changes. In period $t+1$ higher spot values cause very small increases in the official price of crude, $P^c_{t+1}$, and consumer prices of petroleum products. $P^*_{t+1}$, while inducing large changes in the demand for inventories, $I_{t+1}$. At the same time, higher consumer prices cause very small declines in $Q^D_{t+1}$. In general, the initial increases in the demand for inventories will exceed the decline in final demand. As a result expected demand will increase still further, tending to drive up $SV_{t+1}$ beyond the increase induced by the lagged adjustment process. The increase in $SV_{t+1}$ then will contribute to higher values of $P^c_{t+2}$, $P^*_{t+2}$, $I_{t+2}$, and $SV_{t+2}$, while exerting more downward pressure on $Q^D_{t+2}$. The process of tightening will continue until one of two things happens: (1) Either consumer demand plus changes in the demand for inventories can decline to the point where expected demand $(Q^D_{t+i})$ is equal to or less than the level or supply; or (2) the disruption can end or an alternative source of production can be found so that $Q_{t+i}$ exceeds $Q^e_{t+i}$. It must be noted that it is the second action and not the first that has brought about a return to equilibrium in the six episodes we have studied.

It also should be noted that a temporary return to equilibrium between actual and expected production will not immediately restabilize the market because $P^c_t$, $P^*_t$, $Q^D_t$, and $I_t$ will continue to adjust for a period of time. Thus, unless compensating cuts are made in the quantity of oil supplied to maintain balance between supply and expected demand, a surplus will be created that will tend to drive prices down.

In order to demonstrate this process, the simple model of the petroleum market was implemented using parameters derived for monthly periods. The parameters in the model are shown in Table 6-3. Some were derived by reestimation of the structural equations described above, while others were derived mathematically from the parameters estimated from quarterly data. For instance, since worldwide inventory data are not available on a monthly basis, the parameters of the quarterly inventory demand equation had to be

---

2. Note that periods are defined here as months and that the sudden loss in supply does not immediately trigger an increase in the demand for inventories. This may be somewhat unrealistic. However, the model was deliberately specified in this fashion to avoid a simultaneous equation problem between the inventory demand equation and the spot value equation. The recursivity of the model makes it easier to describe and work with the model.

Table 6-3. Parameters Used in Model of Oil Markets.

---

Equation (6.1). Official price of crude oil

$\alpha = .07$

$(1 - \alpha) = .93$

Equation (6.2). Markup for consumer prices

$A = \$6.00/bbl$

Equation (6.3). Demand for inventories[a]

$\beta_0 = 1,041$ (constant)

$\beta_1 = 84.$ (bbls per day per dollar spread)

$\beta_2 = .82$ (adjustment parameter)

Equation (6.4). Change in spot prices

$\gamma = 2.0$ (elasticity of change in spot price per percentage of shortage)

Equation (6.A1). Consumer demand

$\delta_0 = .341$

$\delta_1 = -.02$

$\delta_3 = .93$

---

a. The monthly inventory equation was obtained by solving the quarterly equation for its steady state value and then solving for a monthly adjustment parameter, which produced results with a monthly equation that were identical to the results from the quarterly equation. Thus, if the quarterly equation was:

$$I_t = \beta_0 + \beta_1 \, SV_{t-1} - (P^c_{t-1} - (r_t \, P^c_{t-1}/(1-\alpha))) + \beta_2 I_{t-1},$$

the monthly version was found by finding a value for $\beta_2$ (holding $\beta_1$ and $\beta_0$ constant) such that the change in stocks predicted by the monthly equation matched the change in stocks predicted from the quarterly equation for any spread maintained for a three-month period.

recalculated to put them on a consistent basis with the other monthly data. The basis of the recalculation is explained in a footnote to Table 6-3.

The completed model provides a mechanism for simulating the movements in official crude prices, spot crude values, levels of inventories, and consumer demand during periods of shortage, glut, or stability, under the assumption that individual consumer countries can be aggregated together. This is a strong assumption employed here. For convenience, the results are reported on a seasonally adjusted basis.[3]

The initial simulations with the model were calibrated to a level of OECD consumption of 40 million barrels a day, and the official price and spot value of crude oil were both set at $30.00 a barrel. The initial simulations of the model (Table 6-4) showed that all characteristics of the market remained stable so long as world production, $Q_t$, remained constant at 40 million barrels a day.

The initial stable state of the model was then perturbed after two months by reducing production by 1 million barrels a day for six months. (This simulation is also shown on Table 6-4.) As the table shows, the reduction in supply quickly triggered an increase in spot crude values from $30.00 to $52.76 a barrel, with the largest increases occurring not during the first month of the disruption but in the last months. The primary explanation for this behavior rests in the structure of the inventory equation.

As may be observed from Table 6-4, the demand for inventories increased. This increase in demand would be expected because the slow response of official crude prices to higher spot values creates an increasing spread between spot values and official prices. This spread in turn induces an increase in the demand for inventories.

The effect of the six-month disruption does not die immediately after supply is restored to its earlier level. In fact, Table 6-4 shows that a mere restoration of supply to the levels prevailing prior to the disruption by no means ends the process of price increases. The cycle of increasing spot values and increasing inventories continues on indefinitely, with prices rising to absurd levels unless some limit on the rate of inventory accumulation is established.

This characteristic of the basic simulation is significant, since, absent some external increase in supply, the price increase continues for a prolonged period. In other words, the model as initially configured is unstable in the short term. This does not necessarily imply that the model is wrong but rather that the oil market might well be

---

3. It was noted above that some of the parameters were derived using data that were not seasonally adjusted. However, statistical tests using dummy variables for each season showed that neither inventory demand nor OPEC prices demonstrated any seasonal variation. This is not surprising because to the extent that seasonal fluctuations are observed they should be caused by seasonal patterns in final demand. That is, seasonal fluctuations in demand induce seasonal fluctuations in OPEC production. Therefore, if parameters for the demand function derived from seasonally adjusted data are employed, the model is effectively put on a seasonally adjusted basis.

unstable if no action were taken to meet a continuous disruption. The cause of the instability lies in the inventory demand equation, because no constraint is imposed on the rate of stock acquisition or on the maximum level of stocks. The section below discusses certain changes that eliminate this problem.

The process of increasing spot values can be stopped if the level of supply at the end of the disruption is established, not at the pre-disruption output level but at a level that equals or exceeds the sum of consumer demand plus the demand for inventories, that is, at or above $Q_t^e$. When production is set at this level, a process of declining spot values begins. This is the effect shown in Part B of Table 6-4.

The cessation of increasing spot values does not bring the process of market disruption to an end. As may be observed from Part B of Table 6-4, increases in official prices continue. These increases gradually suppress the level of consumption and cause declines in the demand for inventories. Thus, unless the level of supply is continually "fine tuned" to reflect these declining demands, a downward spiral in prices is begun.

The initial results highlight the characteristics of the world oil market that prolong and exacerbate disruptions, including the historically sluggish response of official crude oil prices to changes in spot values; the fact (noted in Chapter 4) that consumer prices have historically followed crude prices and not spot values; the speculative nature of inventory demand; and the very sluggish response in consumer demand. The sections below examine the effect that changes in each type of behavior might have in moderating the impact of a disruption.

The initial results also raise an interesting issue concerning the behavior of the members of OPEC. Since prices can be stabilized only if production is increased to a level that satisfies both total consumer demand and the demand for inventories, spot values can begin to decline and may continue to decline unless output is continuously adjusted to stabilize spot values. The questions raised by this behavior are (1) whether exporting countries tend (or have tended) to adjust production in a way that stabilizes spot values while they wait for official prices to catch up or (2) whether they adjust production at the end of a disruption in a fashion that stabilizes official prices?

The answer seems to be that they have in the past acted in both ways. For instance, in the 1973-74 embargo, spot values increased

Table 6-4. Initial Simulation of Oil Model.

| Month | Official Price of Crude | Spot Value of Crude | Change in End of Period Inventory | Quantity Demanded | Shortage (Surplus) |
|---|---|---|---|---|---|
| | Dollars a Barrel | | Million Barrels | Million Barrels a Day | |
| Month 1 | $30.00 | $30.00 | 0.0 | 40.0 | 0.0 |
| Month 2 | 30.00 | 30.00 | 0.0 | 40.0 | 0.0 |
| | | Production Cut by 1 mmbd | | | |
| Month 3 | 30.00 | 31.61 | 0.0 | 40.0 | 1.0 |
| Month 4 | 30.11 | 33.76 | 8.1 | 40.0 | 1.3 |
| Month 5 | 30.37 | 36.61 | 17.4 | 40.0 | 1.6 |
| Month 6 | 30.80 | 40.42 | 28.4 | 40.0 | 1.9 |
| Month 7 | 31.48 | 45.57 | 41.9 | 39.9 | 2.3 |
| Month 8 | 32.46 | 52.76 | 59.4 | 39.9 | 2.9 |
| | | Part A: Production Restored to Previous Level | | | |
| Month 9 | 33.89 | 60.08 | 83.5 | 39.8 | 2.6 |
| Month 10 | 35.72 | 70.24 | 103.1 | 39.7 | 3.1 |
| Month 11 | 38.13 | 85.45 | 132.6 | 39.6 | 4.0 |
| Month 12 | 41.15 | 111.01 | 181.0 | 39.3 | 5.4 |
| Month 13 | 46.32 | 162.84 | 271.2 | 39.1 | 8.1 |
| Month 14 | 54.47 | 309.34 | 474.9 | 38.8 | 14.6 |

*Part B: Production Restored to Level Meeting Expected Demand*

|  |  |  |  |  |
|---|---|---|---|---|
| Month 9  | 33.89 | 52.30 | 83.5  | 39.8 | NA |
| Month 10 | 35.17 | 49.54 | 64.0  | 39.7 | NA |
| Month 11 | 36.18 | 47.03 | 35.9  | 39.6 | NA |
| Month 12 | 36.94 | 42.76 | 13.9  | 39.5 | NA |
| Month 13 | 37.75 | 41.56 | -13.0 | 39.4 | NA |
| Month 14 | 37.64 | 41.31 | -19.0 | 39.2 | NA |

NA: Not Applicable.

to $19.00 a barrel before declining back to $13.00 to $14.00 a barrel; behavior that was similar to that during the 1980 Iran/Iraq war. On the other hand, during various episodes of the 1979 Iranian crisis, it appears that outputs were adjusted periodically in an effort to stabilize spot values.

## THE EFFECT OF VARIATIONS IN THE PARAMETERS OF THE OIL MODEL

The initial results presented in the previous section suggest that the oil market is unstable during a disruption and that even a modest shortage of 1 million barrels a day could cause a prolonged reaction that might result in a destabilizing cycle of price increases and inventory accumulation. The following section investigates the stability of these conclusions by varying the principal equations of the model one at a time in order to determine which of the behavioral relationships can be identified as a principal contributor to the destabilizing behavior observed above. The simulations:

- Vary the speed of adjustment of OPEC prices to spot prices to determine whether more rapid adjustment by OPEC would moderate the effects of the disruption;
- Alter the manner in which consumer prices are determined to investigate whether a system that pegs consumer prices to spot values would increase or decrease the magnitude of the fluctuations;
- Change the description of inventory demand to determine the role of speculative accumulation in crises; and
- Modify the demand functions in several ways to see if quicker adjustment by consumers would make the transition smoother.

In each evaluation the movement of spot product values in the base case is compared with the movement in spot product values in the case at hand. The election of spot product values as the basic means of comparison was made for two reasons: First, it is the most important determinant in the market, and second, the comparison of many different simulations is greatly eased by concentrating on one characteristic rather than several. Finally, there is widespread agreement that the greatest damage resulting from disruptions is caused by the increase in oil prices that accompanies the disruption. In a sense, spot values represent a measure of potential damage.

The simulations presented below represent a means of moving from the discussion of the structure of the market to an assessment of alternative policy options. Thus, while the analysis here is presented in terms of dry statistical results, there is a hidden agenda that will emerge in Part II.

## THE EFFECT OF ALTERNATIVE RATES OF OPEC PRICE ADJUSTMENTS ON MARKET STABILITY

The very long and fundamentally unstable cycle of price adjustments identified in the previous section can easily be made stable within this very simple model by respecifying the adjustment process by which OPEC sets prices. Specifically, the adjustment process that has determined OPEC prices for the last few years is described by the simple adjustment equation discussed in Chapter 2:

$$P_t^c = \alpha P_{t-1}^c + (1 - \alpha) SV_{t-1} . \qquad (6.1')$$

In the monthly model the parameters have the values .93 and .07.[4] This implies that a change of $1.00 in spot values this month will induce only a $0.07 change in official crude prices next month. Given these parameters, a full thirty-three months is required for the official price to capture 90 percent of a permanent increase in spot values. This process can be speeded up dramatically if the parameters of the price equation are adjusted. For instance, the number of months required to recover 90 percent of a price increase with various combinations of parameters of the price equation is shown below on Table 6-5.

---

4. These parameters were estimated using monthly data published in *Petroleum Intelligence Weekly*. The results of the equation (fitted to Arab light prices) are:

$$P_t^c = .931 P_{t-1}^c + .069 SV_{t-1}$$

(.104)

$R^2 = .798$   SEE = 3.33   DW = 0.242

Period of Fit 1978:5 to 1980:12

(Note: No standard error of $1 - \alpha$ is computed).

Table 6-5. Number of Months Required by OPEC to Capture 90 percent of any Increase in Spot Values.

| Value of $\alpha$ | Value of $1 - \alpha$ | Number of Months Required to Capture 90 Percent of Price Increase |
|---|---|---|
| .93 | .07 | 33 |
| .90 | .10 | 23 |
| .85 | .15 | 16 |
| .80 | .20 | 12 |
| .75 | .25 | 10 |
| .70 | .30 | 8 |
| .65 | .35 | 7 |

Table 6-6. Spot Values of Crude Oil Under Different OPEC Pricing Regimes.

| Month | Value of $\alpha$ | | | | | |
|---|---|---|---|---|---|---|
| | .93 | .8 | .6 | .4 | .2 | 0 |
| | Dollars per Barrel | | | | | |
| Month 1 | $30.00 | $30.00 | $30.00 | $30.00 | $30.00 | $30.00 |
| Month 2 | 30.00 | 30.00 | 30.00 | 30.00 | 30.00 | 30.00 |
| | Production Cut by 1 Million Barrels a Day | | | | | |
| Month 3 | 31.61 | 31.61 | 31.61 | 31.61 | 31.61 | 31.61 |
| Month 4 | 33.76 | 33.75 | 33.74 | 33.73 | 33.71 | 33.70 |
| Month 5 | 36.61 | 36.50 | 36.34 | 36.20 | 36.03 | 35.89 |
| Month 6 | 40.42 | 39.96 | 39.34 | 38.80 | 38.30 | 37.95 |
| Month 7 | 45.57 | 44.25 | 42.61 | 41.36 | 40.12 | 39.72 |
| Month 8 | 52.76 | 49.51 | 45.96 | 43.62 | 39.89 | 41.06 |

The effect of introducing more rapid price adjustment by OPEC may be demonstrated by reducing the value of $\alpha$ and increasing the value of $1 - \alpha$. In the extreme, the values would be $\alpha = 0$ and $1 - \alpha = 1.0$. Such a change has the effect of increasing the responsiveness of oil markets as we can see from Table 6-6. As $(1 - \alpha)$ increases, the increase in spot values is much smaller. Furthermore, the basic instability identified in the base case is eliminated when $(1-\alpha)$ is greater than 0.6. This means that the price increases tend to dampen of their own accord even if the shortage is not eliminated. (Recall that in the base case an increase in supply was required to stop the process of price increases.) The obvious conclusion is that the process of price setting used by OPEC tends to make the effects of a disruption worse. Month-to-month changes in spot product values for various values of $\alpha$ and $1-\alpha$ are shown in Table 6-6.

There is a fairly obvious explanation for the fact that more rapid adjustments in world crude oil prices would speed the process of adjustment. That explanation is found in the inventory demand equation. Specifically, changes in the demand for inventories are determined by the spread between spot product values and the cost of crude oil. The more rapid the OPEC price adjustment, the smaller the spread; and the smaller the spread, the smaller the increase in the demand for inventories. Thus, the speed-up in the price adjustment process reduces excess demand and therefore the rate of price escalation.

The more rapid adjustment in world oil prices also speeds up the process of consumer adjustment to higher prices. Due to the parameters of the demand function, however, the effect is much less noticeable.

## THE EFFECT OF ALTERNATIVE MEANS OF SETTING CONSUMER PRICES ON MARKET STABILITY

In Chapter 4 it was noted that consumer prices for petroleum products have tended to follow official crude prices rather than prices prevailing on commodity markets and that this tends to exacerbate the effect of interruptions in the supply of oil because the amount of price-induced conservation would tend to be reduced. To determine the effectiveness of such a change in moderating the effect of a disruption, the model was simulated with a different consumer price determination equation. Instead of using the equation shown in Table 6-1, $P^* = K + P^c_{t-1}$, an alternative equation was used:

$$P^*_t = K + SV_{t-1} . \tag{6.2'}$$

Table 6-7. Impact of Substituting Consumer Prices Based upon Spot Product Values for Consumer Prices Based upon Official Crude Prices.

| Month | Spot Product Values in Base Case | Spot Product Values With Changed Consumer Pricing System |
|---|---|---|
| | Dollars per Barrel | |
| Month 1 | $30.00 | $30.00 |
| Month 2 | 30.00 | 30.00 |
| | Production Cut of 1 Million Barrels a Day | |
| Month 3 | 31.61 | 31.61 |
| Month 4 | 33.76 | 33.20 |
| Month 5 | 36.61 | 36.35 |
| Month 6 | 40.42 | 39.63 |
| Month 7 | 45.57 | 43.65 |
| Month 8 | 52.76 | 48.53 |
| Month 9 | 63.29 | 54.48 |

Under this hypothesis, consumer prices were assumed to follow spot prices. The effect of this change on the behavior of the demand for inventories, changes in spot values, and consumer demand was small but noticeable. (See Table 6-7.) However, the shift in pricing regimes does not, by itself, eliminate the basic instability of the model identified above. Just as in the initial simulations, stability can be restored only if production is returned to a level that is substantially in excess of the original level of production.

However, the effect of a change in pricing regimes may introduce much more fundamental changes in the behavior of the market by affecting the inventory decision process. This would happen if the demand for inventories were affected by a change in pricing policies, and such a change should be expected. In fact, the entire speculative justification for acquiring inventories at the start of a disruption would be eliminated because the potential profit realizable by purchasing crude oil while crude prices were below spot values and then holding the crude in inventories until prices caught up with spot values would be eliminated. Thus, under a regime where consumer prices rose with spot values, the speculative motive for holding inventories would be eliminated.

The effect of this type of change is shown in Table 6-8. As may be observed, the consequences are much more dramatic than those shown in Table 6-7. Increases in spot values and official OPEC prices are much smaller, and the combination of setting prices at spot values and eliminating the speculative aspect of inventory acquisition stabilizes the market after a disruption without an increase in production.

### The Effect of Alternative Inventory Policies

One of the principal empirical conclusions of this study is that the oil industry acquires inventories when spot values are greater than official crude prices and disposes of stocks when spot values fall below official prices. This argument has been made by many critics of the oil industry in the past and is just as frequently refuted by those in the industry.[5] This section evaluates the effect of changes in purchases and sales from stocks in stabilizing markets and the effect of the equilibrium level of stocks in dampening or exacerbating cycles.

The first set of simulations evaluates model behavior for a disruption that occurs at a time when inventories are either below or in excess of equilibrium levels. This is effectively a simulation of market behavior in 1979 and 1981. (The general impression is that world oil markets were caught with low inventories in 1979 but benefitted from a high level of inventories in 1981 when war broke out between Iran and Iraq.)

The simulations were constructed by setting starting stocks at 97.5 percent of equilibrium level, 100 percent of equilibrium level, and 102.5 percent of equilibrium level at the start of the disruption.

---

5. One senior official of a major oil company stated flatly, "I never once looked at the spot values of oil when deciding whether to add to or liquidate inventories." However, when asked why the company had at one point reduced liftings from a particular African country, the official replied, "Because the netbacks were less than the official selling price."

Many officials of oil companies seem to hold the same view. Spot values (or netbacks) do not affect their decisions to sell from stocks but do affect their willingness to acquire additional supplies. Of course, if they take their sales as a given and vary the amount of oil acquired depending on the difference between spot values and official prices, then they are behaving precisely as this model predicts they should. As a postscript it may be noted that oil companies behaved in precisely this fashion during the first months of 1982.

**Table 6-8.** Impact of Basing Consumer Prices on Spot Values in Lieu of Official Crude Prices, Assuming that Spot Market Pricing Negates Incentive to Build Stocks.

| Month | Base Case Spot Values | Simulation | | | | |
|---|---|---|---|---|---|---|
| | | Spot Value | Official Price | Change in Stocks | Quantity Demanded | "Shortage" (or Surplus) |
| | | $/bbl | | Million bbl | bbl/Day | bbl/Day |
| Month 1 | $30.00 | $30.00 | $30.00 | 0.0 | 40.00 | 0.0 |
| Month 2 | 30.00 | 30.00 | 30.00 | 0.0 | 40.0 | 0.0 |
| | | *Production Cut of 1 Million Barrels a Day* | | | | |
| Month 3 | 31.61 | 31.61 | 30.00 | 0.0 | 40.0 | 1.0 |
| Month 4 | 33.76 | 33.24 | 30.11 | 0.0 | 39.9 | .9 |
| Month 5 | 36.61 | 34.84 | 30.33 | 0.0 | 39.9 | .9 |
| Month 6 | 40.42 | 36.33 | 30.65 | 0.0 | 39.8 | .8 |
| Month 7 | 45.57 | 37.67 | 31.04 | 0.0 | 39.7 | .7 |
| Month 8 | 52.76 | 38.79 | 31.51 | 0.0 | 39.6 | .6 |
| Month 9 | 63.29 | 39.65 | 32.02 | 0.0 | 39.4 | .4 |
| Month 10 | 79.96 | 40.22 | 32.55 | 0.0 | 39.3 | .3 |
| Month 11 | 109.86 | 40.48 | 33.09 | 0.0 | 39.1 | .1 |
| Month 12 | 176.22 | 40.45 | 33.61 | 0.0 | 38.9 | 0.0 |
| Month 13 | 396.89 | 40.15 | 34.09 | 0.0 | 38.9 | -1.0 |

**Table 6-9.** Impact on Spot Values of Initial Stock Levels That Are Above or Below Equilibrium Levels at the Time of a Disruption (*dollars per barrel*).

|         | Stock Level at Start | | |
|---------|------------------|----------------|------------------|
| Month   | Below Equilibrium | At Equilibrium | Above Equilibrium |
| Month 1 | $30.82 | $30.00 | $29.19 |
| Month 2 | 31.78  | 30.00  | 28.30  |
| Month 3 | 34.65  | 31.61  | 28.81  |
| Month 4 | 38.49  | 33.76  | 29.62  |
| Month 5 | 43.72  | 36.61  | 30.79  |
| Month 6 | 51.05  | 40.42  | 32.39  |
| Month 7 | 61.85  | 45.57  | 34.54  |
| Month 8 | 79.08  | 52.76  | 37.38  |

Note: Stocks were set at 97.5 percent, 100 percent, and 102.5 percent of equilibrium level.

Then a loss in production of 1 million barrels a day was imposed in the third month and allowed to extend until the eighth month. The results shown on Table 6-9 indicate that the level of stocks at the start of a disruption will play a critical role in the determination of the movement of prices during the disruption. For instance, in the simulation where stocks were allowed to fall below the equilibrium level, spot values increased to $79.08 by the eighth month (six months after the start of the disruption), while spot values only increased to $52.76 if stocks started in equilibrium. Furthermore, if stocks were above equilibrium level, spot values only increased to $37.38.

The explanation for this behavior is found in the structure of the inventory demand equation. This equation is written as

$$I_t = \beta_o + \beta_1 (SV_{t-1} - (P^c_{t-1} - r_t P^c_{t-1}/(1-\alpha)) + \beta_2 I_{t-1} . \quad (6.3')$$

The equilibrium level of stocks, $I^*$, may be found by solving the difference equation,

$$I^* = \frac{\beta_o}{1 - \beta_2} . \quad (6.A4)$$

Under normal circumstances the level of inventories will tend to gravitate towards $I^*$: When the level of inventories exceeds $I^*$, stocks

Table 6-10. Comparison of Spot Values Under Alternative Inventory Acquisition Assumptions (*dollars per barrel*).

| Month | Base Case | Spot Values Have No Effect on Inventories[a] | Contrary Case (buy high/sell low)[b] |
|---|---|---|---|
| Month 1 | $30.00 | $30.00 | $30.00 |
| Month 2 | 30.00 | 30.00 | 30.00 |
| *Production Cut of 1 Million Barrels a Day* | | | |
| Month 3 | 31.61 | 31.61 | 31.61 |
| Month 4 | 33.76 | 33.53 | 32.83 |
| Month 5 | 36.61 | 35.79 | 33.82 |
| Month 6 | 40.42 | 38.46 | 34.67 |

a. Assumes spread between spot values and official sales prices does not affect inventory acquisition.

b. Assumes oil companies acquire stocks when spot values are below official prices and sell when spot values are above official values.

will be drawn down, while when the level is below $I^*$, stocks will be rebuilt. The actual level of stocks will be moved away from $I^*$ when spot and crude values deviate from one another—as they do during a disruption. The impact of a disruption on the demand for stocks will therefore be moderated if a disruption occurs at a time when stocks are being liquidated because they exceed $I^*$. On the other hand, if a disruption occurs at a time when stocks are being rebuilt, the rebuilding will compound the pressure on demand created by the price spiral.

Two changes in the model were also simulated. In the first, it was assumed the demand for inventories was not influenced by differences between spot values and official crude prices. In the second case, it is assumed that inventories are acquired when official crude prices exceed spot values and sold when crude prices are less than official prices. (This is a variant on the saying "Buy high, sell low.") The results of both sets of simulations are shown on Table 6-10.

In the first case it is assumed that the price spread does not matter. This assumption does not eliminate the basic instability of the model to a disruption. However, the rate at which spot crude values and official prices rise is slower. This first case is equivalent to a deletion of the admittedly poor inventory equation from the model.

The first simulation does indicate, however, that a return to former levels of output is sufficient to quickly reverse the trend in prices. It is no longer necessary, as it was in the base case, to flood the market with oil to restore stability.

The simulation of contrary inventory acquisition policies offers even better results. The sales from stocks reduce the effect of a cut in supply and substantially moderate increases in prices. As a consequence, the increase in spot values during the first six months in response to the loss of 1 million barrels a day is only $6.00 a barrel, or approximately a 20 percent increase in price.

### Changes in Consumer Demand

Three final sensitivity changes were directed at the parameters of consumer demand: (1) reductions in the level of consumption by a direct adjustment in the constant in the demand equation; (2) increases in price elasticity of demand; and (3) increases in the rate of adjustment. Each change was implemented twice. In the first simulation the change was implemented coincidently with the beginning of a disruption. In the second simulation it was implemented three months after the start of the disruption.

Of the six changes that were tried, two had substantial effects on the rate of increases in spot values, and four had almost no effect. The two that were effective were the adjustments in the constants in the demand function. Algebraically, this change was represented by

$$Q_t^D = \delta_0 + \delta_1 \ln(P_t^*) + \delta_2 \ln(Y_t) + \delta_3 \ln(Q_{t-1}^D) - X , \qquad (6.A1')$$

where $X$ represents the amount of consumption that is arbitrarily deducted as a reduction. This type of adjustment might be classified as mandated emergency conservation. In the simulations, arbitrary reductions in consumption that were equal to the cut in production tended to fully offset the cut in production. Indeed, as may be observed from Table 6-11, the cut in consumption tended to drive spot values below their initial level. The explanation for this anomalous behavior rests in the dynamic structure of the demand function. In these simulations, an initial cut in consumption has an increasing effect over time, just as an increase in price or decline in real income will have greater long-run than short-run effects.

Table 6-11. Comparison of Spot Values of Crude Oil Under Different Parameters of the Demand Function.

Spot Values of Crude Oil (dollars per barrel)

| Month | Base Case | Constant Change | | Price Elasticity Change | | Adjustment Parameter Change | |
|---|---|---|---|---|---|---|---|
| | | Immediate | Three-Month Lag | Immediate | | Immediate | |
| Month 1 | $30.00 | $30.00 | $30.00 | $30.00 | | $30.00 | |
| Month 2 | 30.00 | 30.00 | 30.00 | 30.00 | | 30.00 | |
| Month 3 | 31.61 | 30.00 | 31.61 | 31.61 | | 31.61 | |
| Month 4 | 33.76 | 28.54 | 33.76 | 33.76 | | 33.76 | |
| Month 5 | 36.61 | 25.55 | 36.61 | 36.60 | | 36.61 | |
| Month 6 | 40.42 | 21.30 | 38.41 | 40.37 | | 40.42 | |
| Month 7 | 45.57 | 16.43 | 38.56 | 45.44 | | 45.58 | |
| Month 8 | 52.76 | 11.75 | 36.52 | 52.42 | | 52.76 | |

Lump-sum cuts in consumption of this sort also appear to have beneficial effects on spot values, whether they are imposed immediately or after the crisis has begun. In fact, when imposition of the 1 million barrel a day cut in consumption was delayed for three months, it was still effective in quickly stopping the process of price escalation.

Experiments with changes in the price elasticity of demand and changes in the rate of adjustment parameter of the demand function were not successful. In fact, an almost negligible effect was observed, even when the changes in these parameters were made at the start of the disruption. This suggests that the absolute magnitude of the short- and long-run price elasticity of demand is relatively unimportant in determining the effectiveness of measures for meeting disruptions.[6] This conclusion should not be surprising, since the price that affects consumer demand is the consumer price and not the spot value. Since the consumer price tends to move with the official price of crude, not its spot value. Thus, changes in price elasticity will have little impact on the rate of change in spot values resulting from a loss in supply, unless the process by which consumer prices are set is altered.

## CONCLUSIONS

The results of simulation of this small model of short-run behavior of oil markets allow us to draw a number of conclusions as to the behavior of these markets when faced with relatively minor disruptions. First, the impact of a loss in supply on changes in prices may last long after supply has been restored. The explanation for this phenomenon rests in the character of the demand for inventories. When spot values of crude oil exceed the cost of the crude, an incentive to increase stocks is created. This incentive increases demands on oil markets and tends to bid up prices unless the volume of supply after the disruption is temporarily greater than the volume of supply prior to the disruption.

---

6. The short- and long-run elasticities of demand are related by the formula:

Long-run elasticity = Short run-elasticity/ (1 – adjustment parameter).

Since changes in either the adjustment parameter or the short-run price elasticity essentially affect the long-run elasticity, the discussion here is in terms of the short- and long-run elasticity, not the price elasticity of demand and the adjustment parameter.

Second, the lethargic manner in which crude oil prices are adjusted appears to contribute to the severity of the disruption. When the process of adjustment is accelerated, the rise in spot values is reduced and the period of the disruption is shortened. Furthermore, consumers will benefit from lower long-run prices to the extent that the volume of supply after the disruption is adjusted by OPEC to stabilize spot values at the higher level and thus align crude oil prices with the new, higher, price base.

Third, changes in policies affecting the determination of consumer prices will not, by themselves, have a great impact in modifying the severity of the disruption. To the extent that there is a speculative demand for oil, however, a change in the method of setting prices that causes consumer prices to rise dollar-for-dollar with spot values will eliminate the speculative incentive. The elimination of this incentive will, in turn, pay a substantial dividend in terms of reducing the escalation of spot values. Indeed, such measures by themselves converted potentially unstable simulations into stable ones.

Fourth, the level of inventories at the onset of the crisis is a critical determinant of the behavior of prices during and after the disruption. If a disruption begins at a time when stocks are deemed to be excessive, the impact on spot values and consumer prices will be relatively mild. On the other hand, if the crisis begins when inventories are low, the disruption may have a catastrophic impact on spot values and, if OPEC elects to follow the spot values, disastrous impacts on the economies of consumer countries.

Fifth, the slow adjustment of official crude prices creates a spread between the cost of crude and the spot value that causes firms to increase their stocks of crude during a disruption. These acquisitions tend to destabilize the oil market. The behavior of spot values during a disruption becomes much less volatile when the inventory acquisition practices are stopped or reversed.

Sixth, changes in the characteristics of the demand for oil have little effect in moderating the impact of a disruption unless they affect the immediate level of consumption. Thus, changes that work on or through the price elasticity of demand have minimal impact, even if consumer prices increase quickly, while actions that cut the immediate level of consumption can dramatically slow or even reverse the impact of increased spot values. Furthermore, absent changes in consumer prices, price elasticities are unimportant.

These conclusions are drawn from simulations of a small disruption through the model of oil markets, but simulation of large disruptions produces similar conclusions. Indeed, the basic conclusion to be drawn from the simulations of larger disruptions is that the changes in the structure of the market that were found to be beneficial in the simulations discussed in this chapter are even more important in moderating the effect of a large disruption.

## POSTSCRIPT

The accuracy of an economic model may be determined by measuring its ability to simulate history. Thus, a summary is offered here of model simulations of oil markets during two periods of crisis: the first beginning in October 1978 and extending to June 1979 and the second beginning in September 1979 and extending through May 1980. While the simulations suggest that the model worked reasonably well, great emphasis should not be placed on them for three reasons. First, the measurement of the production shortfall during a disruption is subject to great uncertainties. Second, there is a strong psychological element that affects perceptions of shortages, and this perception has a clearly observable impact on the market. Third, there is a strong element of corporate statesmanship or government pressure in the management of inventories during each crisis, an element which has been ignored in this model.

The effect of uncertainty about the volume of crude production and its effects on spot values can be observed most clearly by examining the behavior of simulated and actual spot values during the Iranian disruption, particularly in February 1979, when Saudi Arabia announced a drastic production cutback. While this cutback was not fully implemented, the effect of the announcement was to drive spot values of Arab light up by a full $7.45 a barrel. However, the model simulated only a $3.11 increase in prices because the cutback was not fully implemented.[7]

The Iranian episode also emphasizes the difficulty, indeed the almost impossibility, of defining the shortfall in expected production. For instance, in November and December there was a fairly clear shortfall in output which we put at 500,000 barrels a day and 2 million barrels a day, respectively. (The loss in Iranian production

---

7. Recall that the variable that causes spot values to move is given by the ratio of expected production (which equals expected demand plus desired change in stocks) to actual production. In February 1979 the Saudi announcement had the effect of forecasting a value of production for February that later was proven to be incorrect. The movement in prices in February appears to respond to the forecasted level of production rather than the actual level of production.

was greater, but some of the loss was made up by increased production from Saudi Arabia and other exporting countries.) For the month of January no loss in production is assumed, even though Iranian production almost ceased for most of the month. This assumption relates to the seasonal pattern of oil output (normally a decline in output of 3 to 4 million barrels a day would have been expected in January) coupled with the continued willingness of Saudi Arabia to produce almost 10 million barrels a day. The assumption is, however, open to serious challenge. For instance, materials cited above indicate that some in the U.S. Department of Energy were quoted by *Petroleum Intelligence Weekly* in late January as still clearly worried about a shortage.

The assessment of expected production from February through June is somewhat easier. In February the shortfall was on the order of 1 million barrels a day, although, as noted above, it appeared that the shortfall might be closer to 2 million barrels a day at the start of the month. Then in March and April the simple seasonal adjustment of the data described in Chapter 2 suggest that production was perhaps 500,000 barrels a day less than demand. In May production probably equaled expected demand. Finally, in late May a second cut in Saudi production created a shortfall of 800,000 barrels a day and once again drove prices up.

Each of these assessments is open to challenge, and none can be measured with great accuracy. However, when they are fed into the model of oil market behavior, the predicted spot values appear to follow the trend of movements of the actual data, with the one clear exception of the February production cutback. The predicted and actual prices are shown on Table 6-12.

Predicted official prices of OPEC crude also appear to track actual prices through March, when a gap of $2.00 to $3.00 develops. This can be explained in part, however, by the model's missing one step in the movement of spot values.

An assessment of the model's success in tracking the movements in inventories is more difficult because data on the level of stocks in the OECD area are available only on a quarterly basis. However, from the data that are available, it must be noted that the model predicted that stocks would be increased by 22.7 million barrels in the first quarter, when actually 40 million barrels of stocks were liquidated.

The model was more accurate in the second quarter. Stocks were predicted to increase by 26.7 million barrels and actually increased

**Table 6-12.** Model Performance in Assessing Movement of Spot Values and Official Crude Prices During the Iranian Crisis, November 1978 to June 1979.

| Month | Spot Values (dollars per barrel) | | | Official Prices[a] (dollars per barrel) | | |
|---|---|---|---|---|---|---|
| | Predicted | Actual | Error | Predicted | Actual | Error |
| October, 1978 | $15.36 | $15.36 | $0.00 | $12.70 | $12.70 | $0.00 |
| November, 1978 | 16.25 | 12.40 | -1.15 | 12.86 | 12.70 | .16 |
| December, 1978 | 18.45 | 16.68 | 1.77 | 13.03 | 12.30 | .33 |
| January, 1979 | 19.31 | 19.44 | -.13 | 13.02 | 13.34 | -.32 |
| February, 1979 | 22.42 | 26.89 | -4.47 | 13.54 | 13.48 | .06 |
| March, 1979 | 23.13 | 25.74 | -2.61 | 14.16 | 13.98 | .18 |
| April, 1979 | 23.57 | 25.96 | -2.39 | 14.29 | 17.25 | -2.46 |
| May, 1979 | 27.52 | 30.06 | -2.54 | 15.40 | 18.14 | -2.70 |
| June, 1979 | 33.44 | 35.37 | -1.93 | 16.25 | 19.68 | -3.43 |

a. Arab light.

by 24.5 million barrels, according to the OECD data. While the inaccuracy in the first quarter is disheartening and the accuracy in the second quarter is encouraging, little confidence should be placed in either estimate because levels of inventories were being influenced by several external events.

First, oil companies were at times under extensive pressure to build stocks while under pressures to liquidate them at other times. Second, the seasonal swings in consumption have an effect on first-quarter stocks. For instance, Danielson (1979) argues that the normal seasonal reduction should be 270 million barrels. If this is the case (and we note that we were unable to determine that there was a significant statistical pattern to inventories), then the 40 million barrel actual reduction noted in the first quarter would translate into a 210 million barrel increase on a seasonally adjusted basis.

The behavior of inventories during the Iran/Iraq conflict presented a different problem in constructing historical simulations, because the level of stocks at the start of the conflict was generally admitted to be excessive and because there were heavy public pressures on oil companies to use stocks during the crisis. The simulations were con-

**Table 6-13.** Comparison by Actual and Predicted Spot and Official Prices During the Iran/Iraq War.

| Month | Spot Values | | Official Prices (Arab Light) | |
|---|---|---|---|---|
| | Actual | Predicted | Actual | Predicted |
| | Dollars per Barrel | | | |
| September, 1980 | $31.16 | $31.16 | $30.70 | $30.20 |
| October, 1980 | 34.78 | 34.11 | 30.00 | 30.70 |
| November, 1980 | 38.39 | 37.62 | 32.00 | 31.04 |
| December, 1980 | 36.03 | 37.59 | 32.00 | 31.33 |
| January, 1981 | 36.40 | 35.59 | 32.00 | 34.80 |
| February, 1981 | 35.10 | 34.95 | 32.00 | 31.86 |
| March, 1981 | 35.35 | 34.02 | 32.00 | 32.03 |

Note: Actual spot values calculated by the author from *Platt's* data. Actual official prices from PIW.

**Table 6-14.** Comparison of Predicted Spot and Official Price, Assuming that Inventories Are High at the Start of the Iran/Iraq War with the Case Where Inventories Were Low.

| Month | Spot Values | | Official Prices | |
|---|---|---|---|---|
| | High Inventory | Low Inventory | High Inventory | Low Inventory |
| | Dollars per Barrel | | | |
| September, 1980 | $31.16 | $31.16 | $30.20 | $30.70 |
| October, 1980 | 34.11 | 38.01 | 30.70 | 30.70 |
| November, 1980 | 37.62 | 47.52 | 31.04 | 31.07 |
| December, 1980 | 37.59 | 52.89 | 31.33 | 32.15 |
| January, 1981 | 35.59 | 59.75 | 34.80 | 33.54 |
| February, 1981 | 34.95 | 68.15 | 31.86 | 35.28 |
| March, 1981 | 34.02 | 79.86 | 32.03 | 37.49 |

Note: In the high case the actual level of inventories was used. In the low case inventories were assumed to be at the level experienced at the start of the Iranian crisis.

structed on the assumptions that the starting level of stocks was too high and that no stock building took place during the crisis.

In addition, the cut in production due to the crisis was assumed to amount to 2.7 million barrels a day in October 1980 (that is, expected production exceeded actual production by 2.7 million barrels a day) and the cut in production in November was assumed to amount to 2.3 million barrels a day. After November, Saudi Arabia and the other members of the cartel were assumed to be willing to offset the loss in production, so that in December and January there was no shortfall. Finally, in February and March actual output was assumed to exceed expected (or required) output by 1 million barrels a day as Saudi Arabia attempted to glut the market.

The results of the simulation, shown on Table 6-13, indicate that the model closely tracks the movement of spot values. In fact, the performance appears to be so accurate that we are forced to again caution that the results are highly dependent upon the subjective choice of the expected production.

As a final remark, it should be noted that the world's oil consumers were lucky that inventories were high at the start of the war and that firms elected to liquidate, rather than rebuild, inventories during the crisis. As a measure of that luck, Table 6-14 offers a comparison of simulated spot values under an assumption that inventories were high at the start of the war with a simulation conducted under the assumption that inventories were low at the start of the war. The effect on spot crude values is striking. According to the simulations, spot crude values would have risen to almost $80.00 a barrel by March 1981 instead of only $34.00 a barrel, even if oil production had returned to normal levels by the end of December 1980. In other words, the world escaped from another episode of "price ratcheting" only because inventories were high.

# PART II

# 7 THE EFFECTIVENESS OF PRICE CONTROLS DURING A DISRUPTION

> To improve our lot, we must see the world without illusions, as it really is, like it or not—and however uncomfortable that may be for timid politicians and mindless pollwatchers.
>
> W. Michael Blumenthal (1979: 2)

The price controls in effect in the United States during all six disruptions of the 1970s severely limited or eliminated the refiner's flexibility to increase prices of petroleum products or crude oil. Some observers have claimed that consumers benefitted from these programs because price increases were smaller than they would have been in the absence of controls. However, as is shown here, these benefits were extremely short-lived because price controls both delay and magnify the eventual increase in prices.

## THE FORM OF PRICE CONTROLS

The U.S. price controls program was based on raw materials costs. Firms generally were granted the flexibility to increase prices when the cost of raw materials increased but were prevented or discouraged from raising margins. In the case of oil, controls were established at three levels. First, the prices of a portion of the raw material feedstocks were controlled so that ceiling prices of some volumes of crude oil could be no higher than $6.00 a barrel when imported oil was selling for more than $30.00 a barrel. Second, the margins on certain products were controlled so that their prices could be no higher than the cost of crude oil plus a specific amount. Finally, a third level of controls was imposed on retailers that limited the price that could be charged for a specific product to the price paid for the product plus some limited margin.

At times a fourth level of controls was also imposed on the oil industry in the form of gross margin limitations. These restrictions did not prohibit a firm from increasing the prices of a particular good but limited increases in all product prices by placing ceilings on profits.

Many have argued that the U.S. consumer benefitted from these types of controls during the 1973 embargo and the 1979 Iranian interruption. For instance, Owens (1974: 1313) argues,

> A comparison of wholesale price movements in the U.S. and abroad indicates how much less the country was subject to drastic swings in the prices of major products. One advantage was that, relative to most other industrial countries of the world, the U.S. is primarily dependent for its oil supply on indigenous production and the bulk of that production was price controlled under Phase IV and the two tier system. The other advantage was that the cost pass-through system allowed purchasers of incremental product to compete in the world market and yet essentially made certain that they would not reap large windfall profits as a consequence.

This view was echoed by members of Congress and many representatives of consumer and labor groups. Indeed, as late as 1981 President Reagan was strongly criticized for removing all price controls from petroleum products.

Did consumers realize any longer term benefit from price controls? This question has been addressed by Phelps and Smith (1977) and by Kalt (1980). In general, they conclude that consumers received few benefits but base their analyses on the assumption that world oil prices were predetermined. This chapter seeks to determine whether price controls affected the movement of world oil prices. An analysis based on the model of oil market behavior leads to the conclusion that consumers actually lost.

## QUANTIFICATION OF THE EFFECTS OF PRICE CONTROLS

The quantitative effects of price controls can be assessed within the model of oil market behavior because the model includes an explicit representation of the relationship between consumer prices and world oil prices as well as a representation of the relationship between the demand for inventories and the spread between consumer prices and spot values. The base case specification of this representation assumes that consumer prices adjust with changes in the official price of crude oil for reasons which were spelled out in detail in Chapter 2. In Chapter 6 this specification was written as

$$P_t^* = P_t^c + K, \qquad (7.1)$$

where $P_t^*$ represents the consumer price of petroleum products and $P_t^c$ represents the price of crude. As was noted in Chapter 2, this form of representation presumes that there is some form of price control imposed on refiners that prevents them from adjusting product prices to changes in spot values. Indeed, the specification given in equation (7.1) almost exactly replicates the form of price controls used in the United States.

However, in the U.S. the situation was complicated by the imposition of controls on domestically produced crude oil. Under this regime, the ceiling price for petroleum products would be written as

$$P_t^* = a_1 P_t^{c\prime} + a_2 P_t^{c\prime\prime} + K, \qquad (7.2)$$

where $P_t^{c\prime}$ represents the price of uncontrolled or imported crude oil, $P_t^{c\prime\prime}$ represents the price of domestically produced crude which is under price controls, and $a_1$ and $a_2$ represent the percent of controlled and uncontrolled oil refined.

In the United States the imposition of cost based controls such as are described in equations (7.1) and (7.2) was defended by politicians and government officials on the basis that such controls would reduce the overall rate of inflation and tend to reduce the macroeconomic effects of higher oil prices. In other developed countries, such as France, Japan, and many other members of the Organization for Economic Cooperation and Development (OECD), cost-based controls appear to have been imposed on oil as part of an overall system of administered pricing. The consequence of these various programs was, as we saw in Chapter 3, to cause changes in consumer prices to result from changes in the official price of crude.

To assess the effects of such controls on the behavior of oil markets, we simulated the effects of disruptions through the model of oil market behavior using three different hypotheses about price controls. The goal of these simulations was to determine whether price controls have the effect of reducing or exacerbating the increases in prices experienced during disruptions.

The simulations assume that the price controls program is imposed throughout the OECD, and, as we noted in Chapter 6, that the behavior of consumption in the OECD can be analyzed as a single aggregate entity. The simulations also assume that prior to the disruption an infinite supply of oil is available to the OECD countries at a fixed price and that markets are in an approximate state of equilibrium when the disruption hits. Once the disruption hits, a finite and known quantity of oil is assumed available to consuming countries.

Three types of disruptions are examined. The first is a supply reduction of 1 million barrels a day; the second a reduction of 4 million barrels a day; and the third a reduction of 8 million barrels a day.[1] Each disruption is assumed to last one year. In reality, interruptions of these sizes might occur as a result of the loss or production in a small producing area such as Alaska or Kuwait, in the case of a 1 million barrel a day loss; a loss of output in Iraq or, in the past, Iran in the case of the 4 million barrel a day cut; and the loss of all production from Saudi Arabia or a combination of countries in the case of the 8 million barrel a day cut. The effectiveness of price control programs was assessed by simulating the model of oil market behavior in each disruption under three different types of pricing regimes: aggressive pricing, product price controls, and product plus crude pricing controls.

In the aggressive pricing case, consumer prices were assumed to follow changes on spot markets. Under this condition, any change in the value of crude oil on the spot market in one month is assumed to appear in the consumer price of petroleum products in the next month. The adoption of such a program would imply that oil companies would not wait for the members of OPEC to raise prices but

---

1. These cuts are roughly comparable to the reductions analyzed by the National Petroleum Council Study (1981).

would immediately adjust their prices to reflect movements on spot markets.[2]

The product price control case assumes that the prices of consumer products are adjusted in response to changes in the official world price of crude oil, not its spot value. This type of behavior (described in equation (7.1)) has been mandated by price controllers in many countries. It also may be most representative of the actual practice of administrative price setting followed in the oil industry in the absence of controls.

The product plus crude oil control case assumes that a fraction of the volume of crude oil produced and sold in the consuming countries is subject to price controls and does not increase with world crude prices. This is the type of controls modeled in equation (7.2). Under this system, a $1.00 increase in the price of crude oil purchased from OPEC causes less than a $1.00 increase in the price of products purchased by consumers because the price of a fraction of the crude oil consumed in the OECD does not increase. For instance, a $1.00 increase in OPEC crude causes only a $0.75 increase in products if 25 percent of the crude oil consumed in the OECD is under price controls and does not change in price.

To assess the effects of price controls on the behavior of markets, the simulations were constructed under the assumption that no other form of emergency action was taken during the disruption. Instead, it is assumed that the market is allowed to solve the problem without any further helping hand from the government (except, of course, for the mandated assist in the form of price controls).

The assessment of the effectiveness of controls did, however, also examine the effects of stock levels in modifying price increases during disruptions. This was accomplished by simulating each price control case under two different starting stock levels. In the first, it was assumed that industry stocks at the start of the disruption were 2.5 percent above equilibrium levels. Under these conditions the disruption would occur at a time when industry was attempting to reduce stocks. In the second simulation it was assumed that inventories were 2.5 percent below equilibrium levels, which would mean that the

---

2. There may be reluctance on the part of executives of oil companies to adopt such a pricing strategy for fear that it would cause OPEC to change its pricing strategy. These executives naturally would be reluctant to be blamed for causing OPEC price increases. This practical problem is ignored here.

industry would be attempting to rebuild inventories at the start of the disruption.

In making the simulations, one other assumption concerning inventories was made. During the crisis, the rate of stock acquisition was limited to a maximum rate of 1 million barrels a day. This assumption is required for the two price control simulations to prevent the inventory equation from predicting unrealistic rates of stock acquisition.[3]

The results of the simulations are shown on Tables 7-1 and 7-2. Table 7-1 sumarizes the spot crude values in each case at the end of six months and one year after the start of the disruption. Table 7-2 shows the OPEC price of crude oil six months and one year after the start of the disruption. (Other data on the month-to-month movement of inventories, spot value, and crude prices have been eliminated here to make the presentation manageable.)

These results suggest that price control programs can have a very substantial impact on the increases in spot crude values and official crude prices. For instance, after twelve months of a 4 million barrel a day disruption, the model predicts that the official price of crude oil would have approximately doubled under the aggressive price regime—rising from $30.00 a barrel to roughly $60.00 a barrel, while it could have risen as high as $130.00 if price controls had been imposed.

There are two clear explanations for this result. First, consumer prices increase very rapidly under the aggressive pricing regime, and this increase quickly causes a price induced conservation response. Second, there is little or no speculative demand for inventories in the price responsive simulation because consumer prices rise dollar-for-dollar with spot values. For these two reasons, excess demand is quickly reduced to the level of supply, and the spiral of price increases is cut short.

Despite this explanation, it may be difficult to believe that the price of oil could increase from $30.00 a barrel to over $600.00 in a period of twelve months in the face of a 20 percent cut in supply. Indeed, we would fully anticipate that a government that had the audacity to impose price controls would also impose other emer-

---

3. The rate of stock acquisition is determined by the spread between the spot value and official (or consumer) prices. In the event of a large disruption, this spread may become very large if price controls are in effect. It has the effect of predicting rates of inventory acquisition which are absurd.

Table 7-1. Predicted Spot Crude Values Under Three Different Shortage Scenarios with Different Price Control Regimes.

| Shortage and Inventory Position at Start | Consumer Price Regime | | |
|---|---|---|---|
| | Aggressive | Product Price Controls | Product and Domestic Crude Controls |
| | Dollars per Barrel | | |
| | Spot Value After 6 Months | | |
| **High Inventories** | | | |
| Shortage = 1 mmbd | $ 37.55 | $ 41.35 | $ 47.96 |
| Shortage = 4 mmbd | 85.53 | 129.87 | 144.38 |
| Shortage = 8 mmbd | 276.30 | 526.19 | 588.00 |
| **Low Inventories** | | | |
| Shortage = 1 mmbd | 38.79 | 53.23 | 58.97 |
| Shortage = 4 mmbd | 86.81 | 138.84 | 154.34 |
| Shortage = 8 mmbd | 280.28 | 550.95 | 616.04 |
| | Spot Value After 12 Months | | |
| **High Inventories** | | | |
| Shortage = 1 mmbd | 40.00 | 69.95 | 99.10 |
| Shortage = 4 mmbd | 91.97 | 355.78 | 498.96 |
| Shortage = 8 mmbd | 284.97 | 2,293.71 | 3,297.59 |
| **Low Inventories** | | | |
| Shortage = 1 mmbd | 39.61 | 82.29 | 113.18 |
| Shortage = 4 mmbd | 90.71 | 363.37 | 510.92 |
| Shortage = 8 mmbd | 287.87 | 2,298.79 | 3,311.55 |

Note: Each shortage is assumed to last for a full twelve months. Under the aggressive pricing regime, consumer prices are assumed to move dollar-for-dollar with prices on spot markets. Under the product price control strategy, all crude oil is assumed to sell at the same uniform uncontrolled price. However, price controls are imposed on consumer prices so that consumer prices can change only when crude oil prices are increased. Under the product and domestic crude regime we assume that price controls are imposed on a fraction (25%) of crude oil as they were during the 1970s.

Note that under all simulations, a limit on the rate of inventory accumulation of 1 million barrels a day is imposed.

High inventories are assumed to be 2.5 percent above equilibrium levels, and low inventories are assumed to be 2.5 percent below equilibrium levels.

**Table 7-2.** Predicted Official Crude Prices Under Three Different Shortage Scenarios with Different Price Control Regimes.

| Shortage and Inventory Position at Start | Consumer Price Regime | | |
|---|---|---|---|
| | Aggressive | Product Price Controls | Product and Domestic Crude Controls |
| | *Dollars per Barrel* | | |
| | *Official Crude Prices After 6 Months* | | |
| High Inventories | | | |
| Shortage = 1 mmbd | $ 31.08 | $ 31.24 | $ 31.83 |
| Shortage = 4 mmbd | 38.12 | 41.28 | 42.31 |
| Shortage = 8 mmbd | 59.30 | 72.50 | 75.71 |
| Low Inventories | | | |
| Shortage = 1 mmbd | 31.51 | 33.22 | 33.76 |
| Shortage = 4 mmbd | 38.51 | 42.79 | 43.86 |
| Shortage = 8 mmbd | 60.14 | 75.22 | 78.40 |
| | *Official Crude Prices After 12 Months* | | |
| High Inventories | | | |
| Shortage = 1 mmbd | 34.02 | 39.23 | 44.90 |
| Shortage = 4 mmbd | 58.85 | 109.20 | 130.79 |
| Shortage = 8 mmbd | 160.07 | 542.46 | 686.84 |
| Low Inventories | | | |
| Shortage = 1 mmbd | 34.51 | 45.16 | 50.88 |
| Shortage = 4 mmbd | 59.23 | 114.14 | 136.77 |
| Shortage = 8 mmbd | 160.91 | 556.17 | 704.64 |

Note: The Notes to Table 7-1 also apply to Table 7-2.

gency measures of dubious merit. Furthermore, as prices increased, the price elasticity of demand would increase, dampening the effect on the price of crude. Thus, we offer results, not as forecasts, but as predictions of the outcome of a disruption if only one measure, price controls, were taken at the onset.

The following conclusions may be drawn from Tables 7-1 and 7-2.

1. The effect of any form of contemporaneous price controls is to increase prices by the end of the disruption. If, after the disrup-

tion, OPEC cuts production to maintain prices, the effect is then to induce higher long-term prices.
2. The imposition of price controls on crude oil compounds the problem.
3. The level of inventories at the start of a disruption relative to the equilibrium level will effect the size of price increases experienced during a disruption if the disruption is small.

The major conclusion to be drawn from these simulations concerns the difference in postdisruption official crude oil prices in the aggressive simulation and the two price controls simulations. Official prices in the aggressive simulation are much lower than those at the end of the price controls simulation, implying that price controls cause long-run prices to be higher, not lower.

The explanation for this behavior rests on two factors: inventory demand and consumer demand. Aggressive pricing reduces or eliminates the incentive to acquire inventories during a disruption, thus forcing more oil to be sold to final consumers. This tends to reduce the magnitude of the putative shortage and slows the process of bidding up spot values. Aggressive pricing also induces a much faster response by consumers to the high prices of oil. The reduction in demand also tends to slow the process by which spot values are bid up.

While adoption of aggressive pricing would seem to offer the best mechanism for modifying the effects of a disruption on oil prices, this type of pricing scheme would carry with it certain risks. In particular, it is unlikely that the members of OPEC would sit idly and watch consuming governments or oil companies raise prices during a disruption without any consideration for the cost of crude oil. Indeed, it is far more likely that the members of OPEC would respond by immediately raising their prices.[4] This fear has been expressed by the representatives of many consuming countries. It may also cause any prudent oil company official to eschew such a strategy for fear of being blamed by the press for causing increases in crude prices.

---

4. During 1979 and 1980 the press contained frequent references to statements by Shiek Yamani and other heads of delegations to OPEC who argued that their prices were being deliberately held down to "benefit" the western consuming countries and that it was the responsibility of oil companies to pass the "benefits" of low prices on to consumers. The implied threat in these statements was that if companies became "aggressive," then OPEC would become aggressive.

Should such an outcome be feared? To answer this question the aggressive simulation described above was repeated under the assumption that OPEC responds to aggressive pricing by adjusting official crude prices rapidly. In this case a 50 percent adjustment in official crude prices was assumed to be made in the month following any change in spot values.[5]

The results suggest that retaliation by OPEC would reduce some of the gains achieved by adoption of aggressive pricing. However, consumers are still far better off in the event of a severe disruption under an aggressive pricing regime than under price controls. If OPEC does not retaliate, the official price of crude oil might be $58.85 a barrel twelve months after the start of a 4 million barrel a day disruption, while it might be $98.09 if OPEC did retaliate, $109.20 if product price controls were imposed, and $130.79 if both crude and product price controls were imposed. In other words, the best action (but not the only action) to take at the start of a disruption is to put prices up.

It should be noted that the long run risk to consuming countries of OPEC retaliation must be viewed as small because a price increase could be sustained only if OPEC backed further increases by cuts in production. Further, as was noted in Chapter 6, more rapid price adjustments by OPEC lead to *lower* long run prices.

### Economic Impacts

The argument against putting prices up quickly has always been predicated upon the assumption that the macroeconomic effects of higher prices were undesirable. Time and again macroeconomists have counciled the U.S. Congress and the various administrations against aggressive oil pricing strategies. For instance, Otto Eckstein (1975: 340) advised the Senate Interior Committee,

> I strongly urge this committee to protect the incipient recovery of the economy by averting instant decontrol of old oil prices and working with the administration to achieve a compromise phaseout of controls instead.

This view was echoed by many other economists both in 1975 and again in 1979 when President Carter decontrolled oil prices.

---

5. Recall that the base case model assumes that official crude prices change by 7 percent of any change in official prices in the month following the change in spot values.

These views confuse the macroeconomic problem created by higher oil prices with the economic process by which prices increase. Clearly, higher oil prices have a macroeconomic impact. The transition from a lower to a higher level of oil prices will increase unemployment, temporarily reduce the level of output, and cause a decline in the real level of income in all consuming countries. These results will occur because higher oil prices affect a transfer of income from consuming to producing countries and because higher oil prices make certain vintages of capital goods obsolete. This has been clearly demonstrated by Nordhaus (1980b), Fried and Schultz (1975), and repeatedly by the Council of Economic Advisers (1978). Since the aggressive strategy leads to the lowest increase in crude prices and thus the smallest transfer of income, we argue that it should be preferred to either price control strategy.

## CONCLUSION

This chapter has demonstrated that price controls, far from offering any long-run benefit to consumers, actually lead to higher prices. The first rule for dealing with a disruption, therefore, should be to get prices up, get them up fast, and get them up as high as possible. It also should be noted that the underlying fear that increases in consumer prices might incite OPEC to raise prices even further is misplaced, because OPEC will raise prices anyway. The conclusion is that price controls are not a solution to the problem of disruptions, as is suggested by some, but are part of the problem.

# 8 TAX AND TARIFF POLICIES FOR COPING WITH DISRUPTIONS

> Higher prices are a natural response of a market suddenly faced with an excess of demand over supply. Higher prices provide the incentive for emergency conservation and rapid supply enhancements. In theory, it would be possible to achieve all the needed demand reductions and supply expansions without an increase. But in practice we do not come close to this ideal. So higher prices are an essential feature of any major supply disruption.
>
> The only choice left is selecting the beneficiary of the enormous revenue generated by the higher prices. In the past, by default we elected to make foreign oil producers the primary recipients of higher revenues. We even managed to increase these revenues through our price controls and complex entitlements system.
>
> Prepared statement of Professor William Hogan<br>on "Oil Taxes and Oil Emergencies," in Hearings<br>before the Subcommittee on Taxation and Debt<br>Management of the Senate Committee on<br>Finance, *Special Oil Taxes*, (1980:187).

The analysis in preceding chapters demonstrated that the extent of the price increase resulting from a disruption will depend both on the size of the shortage and on the rapidity with which demand and supply are brought back into balance. Thus, two identical shortages could have very different impacts on the increases in the cost of oil if one were met by a policy that included controls on the prices paid by consumers and the other by a program that sought to aggressively push up consumer prices.

Taxes and tariffs offer one mechanism by which consumer prices can be increased quickly in the event of a shortage. Various types of taxes and tariffs have been proposed for dealing with disruptions, including disruption tariffs, emergency refiner taxes, windfall taxes on oil production, and emergency consumer taxes. Most have approximately the same effect on oil consumption when they succeed in raising prices. However, the effects on oil markets differ. Some of the proposals encourage sales from inventories during crises and cause prices to rise quickly, actions that have been shown to promote price stability, while others tend to discourage inventory building prior to a disruption, discourage sales from inventories during crises, and do not cause immediate price increases.

This chapter discusses the various characteristics of tax programs designed to meet disruptions, by cataloguing the various types of taxes that have been proposed and then by comparing the effects of the alternative forms of taxation in modifying the price impact of disruptions. The discussion also addresses the specific size of the tariff or tax. Although most studies have examined the effect of constant disruption taxes, this study shows that a constant tax is inferior to one that declines over time. The final sections of the chapter address two other topics: the macroeconomic problems created by the imposition of disruption tariffs or taxes and the questions of international response by both consumer and producer nations.

## POSSIBLE FORMS OF DISRUPTION TAXES AND TARIFFS

Five alternative tax or tariff proposals are examined here. These are tariffs on imported crude oil and products, taxes on one or more petroleum products, taxes on refinery output, windfall or severance taxes on crude oil, and a quota/auction.

The tariff on imports of crude oil may be imposed as either an ad valorem or specific levy on every barrel of crude oil imported into the United States. Two effects would be expected from such a tax. First, the price of oil produced within the United States (or within the OECD, if the tax were imposed throughout the OECD) would be expected to increase by the amount of the tariff. Thus, if a tariff of $30.00 per barrel were imposed on every barrel of oil imported, domestic postings would be expected to rise by $30.00 per barrel.[1]

Second, consumption should be reduced as refiners increased consumer prices to recover their higher costs. However, if the tariff were imposed only on imports of crude oil, these price increases could be neutralized by importing products from refineries outside the tariff area. Were the tariff imposed on an OECD-wide basis, this effect would be negligible because there are few refineries located outside the OECD area capable of providing large volumes of products to the OECD market. However, if a tariff were imposed only by a few countries, imports from refineries outside the tariff area could undermine the effect of a tariff imposed only on crude.[2]

The problem of tariff-free imports of products can be circumvented by imposing a tariff on imports of both crude oil and products. Such a tariff, like the tariff on crude oil alone, could be in the form of a specific or ad valorem levy. In either case, it would lift the price of all crude and products in the region imposing the tariff by the full amount of the tariff and cause a reduction in consumption.[3]

As an alternative to the tariff, some have recommended that a tax be imposed on the outputs of refiners and distributors. For instance, Alm, Colglazier, and Kates-Garnick (1981:343) propose that the federal government impose an "emergency windfall profits tax" on refiners. They state, "The tax would be levied on increases over the base mark ups for refiners, wholesalers, and retailers of gasoline." The proceeds of such a tax would then be rebated to consumers. They would apply this tax only to gasoline, presumably imposing price controls on other products, but the concept of the refinery tax could be extended to all products.

---

1. Rowen (1980:39) reports that studies at Stanford indicate that the imposition of a tariff should cause domestic price increases equal to only 70 percent of the tariff. This response is, however, hard to understand because the domestic crude posting usually follows the cost of imported oil.

2. Presumably a tariff imposed on crude only would cause the price of products produced by refiners in the country imposing the tariff to increase by the full amount of the tariff—especially if some refiners used only imported oil. So long as these refiners stayed in operation, all prices should increase by the amount of the tariff (because these refiners are the source of the marginal supply). In this case refiners of domestic oil or importers of product reap a windfall.

3. Since it is assumed that a shortage exists and that the supply curve is temporarily totally inelastic, the full amount of the tariff should be passed on to consumers.

One variant of the refinery tax would involve levying a fixed specific per gallon tax on all refinery output. Refiners would be liable for paying the increased tax whether they increased consumer prices or not. Thus, if the levy were set at $10.00 a barrel, a refiner would be required to pay $10.00 to the U.S. Internal Revenue Service (or to the taxing authority in the refiner's home country) for every barrel of products distributed.

Of these two alternatives, only the second guarantees that prices will be increased. The first alternative is triggered only if refiners or distributors increase the prices of products in excess of the increase in the cost of crude oil. Under most conditions, refiners probably would not take such action because, as noted in Chapter 4, refiners tend to adjust prices only after petroleum exporters raise the posted price of crude oil. If they wait for their costs to increase before raising prices, they incur no tax liability. They cannot, however, escape a per barrel levy. Thus, the refiner response to the per barrel levy would be to increase prices by at least the amount of the tax.

The fourth type of tax would be a direct tax imposed on specific petroleum products, such as an emergency gasoline tax. The fifth type of tax is an emergency severance tax on crude oil. This tax would be imposed on domestically produced crude oil and could be imposed either as a specific per barrel levy or as an excise tax in the style of the present crude oil windfall profits tax.

If the tax were imposed as a specific levy, it might be designed to raise the price of domestically produced crude oil quickly. To illustrate, the tax could be imposed on a fraction of the difference between the world price of crude oil and the spot value of crude oil. Under this scheme, the tax at time $t$ would be imposed on the difference between $SV_t$ and $P_t^c$. Thus, if the tax rate were 70 percent and the difference between spot values and official selling prices of crude oil were $20.00, then every barrel of domestically produced crude oil would pay a levy of $14.00.

The argument for a specific emergency tax of this sort is that it would increase the cost of a portion of the crude oil purchased by refiners at a time of shortage. Due to the shortage conditions, this increase could be passed on to consumers in the form of higher prices and thus would not reduce the refiner's profits. Furthermore, since the tax increased the cost of purchased crude oil, the institutional problems created by the manner in which refiners adjust prices to follow the cost of crude oil would be overcome.

An alternative would be a windfall profits tax shaped in the same manner as the present crude oil windfall profits tax. This tax is really an excise tax on a portion of the price received for crude oil. The general form of the tax is given by

$$t = a(P_s - P_b)(1 - x), \qquad (8.1)$$

where $t$ is the tax per barrel (in dollars), $a$ is the tax rate (in percentages), $P_s$ is the price per barrel at which the oil is sold, $P_b$ is the base price set for the oil, and $x$ is the state severance tax rate. The concept is that the federal government will receive a fixed percentage of the windfall (the amount by which the price received exceeds some base price) from the sale of any barrel of oil. The problem with this form of windfall profits tax is that it relies on the exporters of crude oil to raise their prices because domestic crude prices follow world prices. This would defeat the effectiveness of the tax, because, as was noted in Chapter 3, producers are very slow to raise prices.

The final type of tax proposal would be a quota/auction. Under this scheme, importers of oil would establish an emergency quota on the importation of oil equal to or less than the projected supply of available oil. The rights to import oil would then be auctioned off to the highest bidder. The immediate effect of the auction would be to establish a premium or effective tariff on the cost of imported oil equal to the cost of the quota tickets. This premium should then cause the price of crude oil produced within the quota area to increase to the amount of the quota plus the value of quota tickets. The combined effect of the auction plus the increase in the price of domestically produced crude oil would push up consumer prices. The effect of the quota/auction is, then, identical to that of the tariff on crude and products.

However, as Nordhaus (1980a) notes, there is one major difference between the tariff and the quota/auction. The tariff provides short-run certainty as to price, while the quota/auction provides certainty as to the volume of oil imported. Thus, if world supplies of oil available to the OECD are cut by 8 million barrels a day, the imposition of an OECD-wide quota that reduced the level of imports by 8 million barrels a day would immediately eliminate any excess demand on world crude markets. On the other hand, a tariff of, say, $50.00 a barrel would not guarantee that imports would be reduced by the full 8 million barrels a day. Instead, one would have to wait to observe the movement of supply and demand to determine whether a $50.00 a barrel tariff was too large or too small.

## Qualitative Assessment of the Effects of the Various Tax Proposals

The six alternative tax proposals (and the three variants) can be compared in a number of different dimensions. This section ranks them according to the following criteria:

1. Ability to assure an immediate increase in consumer prices;
2. The effectiveness of the proposal in causing a sharp reduction in consumption at the start of a disruption, before official crude prices start rising;
3. The effectiveness of the proposal in automatically encouraging stock building prior to the disruption and encouraging sales from inventories at the outset of the disruption;
4. The level of assurance one may have that the proposal really achieves a reduction in consumption;
5. The effectiveness of the proposal in moderating price escalation on commodity spot markets; and
6. The effectiveness of the proposal in moderating the increases in world crude oil prices.

A qualitative assessment of the various measures is shown in Table 8-1. According to the rankings, the quota/auction appears to offer the best possibilities for meeting a disruption, followed by the tariff on crude and products, and then by an emergency severance tax on crude oil. Taxes on refiners and on individual products are decidedly inferior, while the windfall profits tax appears to be the worst option.

The quota/auction, tariff options, and specific tax on crude oil all rank high for three reasons. First, they offer the greatest certainty that they will in fact achieve a specific reduction because they all increase the cost of crude to refiners, especially since the price of crude produced inside the OECD should increase after the tariff is imposed by the amount of the tariff. The increase in consumer prices would have the greatest chance of reducing consumption. Furthermore, the tariff and quota reward speculators. Under a tariff or

Table 8-1. Assessment of Effectiveness of Various Tax Proposals in Dealing with a Disruption.

| | Measure | Assures a Reduction in Consumption by Forcing Consumer Prices Up | Encourages Stabilizing Inventory Cycle | Assures that a Specific Import Reduction Target Will Be Hit | Effectiveness in Moderating Price Increases on Commodity Markets | Effectiveness in Moderating Increases in Crude Prices |
|---|---|---|---|---|---|---|
| 1. | Tariff on Crude | Satisfactory | Yes | No | Good | Good |
| 2. | Tariff on Crude and Product | Good | Yes | No | Good | Good |
| 3A. | Refiner Windfall Tax | No | No | No | Little | Poor |
| 3B. | Specific Refiner Tax | Good | No | No | Some | Better than 3A but worse than 1 and 2 |
| 4. | Specific Product Tax | Good | No | No | Some | Better than 3A; worse than 1 or 2 |
| 5A. | Emergency Crude Oil Severance Tax | Good | Yes | No | Good | Good |
| 5B. | Windfall Profit Tax | No | No | No | Little | Poor |
| 6. | Quota/Auction | Good | Yes | Yes | Probably | Good |

quota/auction system, a firm or individual that had built speculative stocks could profit from them. As Nordhaus notes, these programs encourage people to buy cheap and sell dear. He also notes (1980a: 32), "Such behavior is price stabilizing and in the interest of the nation." Finally, these proposals rank high because the stabilizing sales from inventories offer the prospect of reducing increases in spot values of crude oil and thus moderating increases in the official prices of crude.

The quota/auction ranks highest among this group because it also assures that a specific reduction in imports will be achieved. The tariff options do not guarantee this effect and thus are ranked slightly lower. The emergency severance tax or specific tax on crude is ranked below the tariff. While it would raise the cost of crude to refiners and thus push up consumer prices, this tax also would tend to stifle the incremental domestic production of crude oil that normally might be expected during a disruption. Although incremental production has been ignored throughout this study, some potential for increased domestic supply may exist during a disruption.[4]

The specific taxes on products sold by refiners or distributors and the specific product taxes are ranked in the second category. These taxes have the same impact on conservation as the tariff and quota/auction but do not offer any rewards to those who build precautionary stocks when prices are low and then sell from those stocks during disruptions. Thus, these taxes would not encourage stabilizing inventory behavior. Instead, refiners would be encouraged to acquire stocks during the early periods of a disruption before the price of crude oil is put up and to hold on to those inventories until the crisis is passed and the tax is removed. Such behavior would tend to push prices on commodity markets even higher, leading to higher OPEC prices.

The windfall profits tax and the refiner windfall tax rank last among the tax proposals because they do not force firms to raise prices and tend to penalize firms that do raise prices. For instance, under the Alm, Colglazier, and Kates-Garnick (1981) proposal, a refiner would be taxed to the extent that increases in his product prices exceeded increases in the prices paid for crude oil. The tax

---

4. The National Petroleum Council notes that the U.S. has the capacity to produce supplies in the amount of 325,000 barrels a day. It also states that this amount of increased supply could only be obtained if certain regulatory approvals could be achieved. See National Petroleum Council (1981:125).

could be avoided by not increasing prices. Since, as noted in Chapter 4, refiners generally tend to wait for OPEC to raise the price of crude before raising prices anyway, most refiners would escape taxation. However, by escaping taxation they would be preventing prices of petroleum products from rising to the new equilibrium level. This failure, in turn, would prevent the conservation response sought during a disruption. Thus, unlike the other taxes, the Alm, Colglazier, and Kates-Garnick proposal would not assure the immediate reduction in consumption at the start of the disruption.

The refiner tax also would deny speculative profits to those who built up inventories while, on the other hand, offering substantial profits to firms that increase inventories at the start of the disruption and wait for OPEC to increase prices. Last in first out (LIFO) inventory accounting practices would enable these firms to escape most, or all, of the tax. This feature is a decided disadvantage.

The crude oil windfall tax suffers from disadvantages similar to those listed for the refiner windfall tax. No incentive to raise prices is offered because the windfall profits tax prevents the producer from capturing most, if not all, of the increased revenue from any increase in the price of crude oil. Furthermore, since there is no compulsion to increase crude prices until world prices increase, the imposition of a windfall tax probably will have no effect on crude prices. Instead, refiners will wait for the members of OPEC to raise the official price of crude oil before increasing the prices charged to consumers and, while they wait, increase inventories.

Therefore, the windfall profits tax would not cause consumer prices to increase and so would not have any immediate effect on consumption. Nor would it offer speculative profits to those who acquired inventories during times of market stability or those who sold from stocks at the start of a disruption. It would not moderate increases on commodity markets and would have little effect on prices charged by petroleum exporters. In short, it would serve no useful purpose. In summary then, based on this qualitative assessment, the various tax alternatives are ranked as follows:

1. Quota/auction;
2. Tariff on crude and product;
3. Tariff on crude only;
4. Specific tax on domestically produced crude;
5. Specific product tax; and
6. Specific tax on refiner output.

## QUANTITATIVE ASSESSMENT OF TAX OPTIONS

The effectiveness of the three generic types of taxes was examined by simulating the model of oil market behavior. The taxes simulated were a tariff on imports of crude and product, a specific tax on products, and a windfall profits tax on refinery output.

In the simulation of the tariff the model was modified in two ways. First, it was assumed that a fixed tariff was imposed as a temporary measure at the start of the disruption. It was assumed that the size of the tariff was set so that speculators would find that it would be more profitable to sell from stocks during the period the tariff was imposed than to hold them until after the emergency passed. Thus, the inventory equation was modified so that stocks were liquidated during the disruption.[5]

The second modification to the model was to assume that consumer prices increased by the full amount of the tariff. This implies that the price of all domestically produced crude oil increased by the amount of the tariff.

In the specific tax on consumer product simulation it was assumed that consumer prices were increased by the amount of the tax but that the imposition had no effect on inventory behavior. Thus, firms were assumed to acquire inventories during the early phases of the disruption, just as they would in the absence of any program. Indeed,

---

5. This was accomplished by redefining the spread variable in the inventory equation. Absent a tariff, inventories are acquired when firms expect future prices to exceed current prices by more than the cost of holding stocks. Thus, in Chapter 5 it was shown that firms should acquire oil if

$$SV_t - P_t^c \geq \frac{r_t P_t^c + K}{1 - \alpha}$$

because spot values, $SV$, are a good predictor of future crude prices. In this simulation it is assumed that the owner or investor can receive

$$P_t^c + T$$

for oil sold from stocks in period $t$ ($T$ represents the tariff), but expects to receive only $P_{t+j}^c$ at some future time when the tariff is removed. Thus, the acquisition/sales rule becomes

$$SV_t - P_t^c \geq \frac{r_t (P_t^c + T) + K}{1 - \alpha}.$$

In the simulation $P_t^c + T$ was substituted for $P_t^c$.

if the tax is viewed as a temporary measure, the incentive to hoard may be increased.[6] Finally, in simulating the refiner windfall tax, it was assumed that refiners would follow traditional practices—that is, that they would not raise prices until the cost of crude was increased.[7]

Each tax proposal was simulated for disruptions of 1 million barrels a day and 4 million barrels a day. In each simulation inventories were assumed to be 2.5 percent above equilibrium levels at the start of the disruption and were assumed to be reduced as the disruption progressed. In the case of the 1 million barrel a day disruption, it was assumed that a tax or tariff of $10.00 a barrel was imposed at the start. In the case of the 4 million barrel a day disruption, it was assumed that a tax or tariff of $40.00 a barrel was imposed.

The results of the simulations are shown on Table 8-2. From these results we observe immediately that the $40.00 tariff has the anomalous effect of driving spot prices down in the case of the 4 million barrel a day shortfall. This occurs because the tariff induces large sales from inventories during the first months of the shortage, totally offsetting the reduction in supply. For instance, in the first month of the disruption the model predicts that refiners would liquidate stocks at the implausably high rate of almost 6 million barrels a day if the market for the oil existed. However, the decline in consumption resulting from the tariff is such that consumers would be willing to absorb only 3.4 million barrels a day.

The straight product tax and the windfall tax on refiners offer no incentive to liquidate stocks, and thus do not tend to modify

---

6. The explanation here is that the tax reduces the value of a barrel of oil in stocks and the value of oil purchased. A firm selling oil to the consumer at $P^*$ receives $P^* - T$ where $T$ is the amount of the tax. Thus, if product price is based upon crude cost, product prices would be set as

$$P_t^* = P_{t-1}^c + T.$$

So long as $SV_t > P_t^*$, the firm should expect $P_{t+1}^c > P_t^c$ and should increase stocks so long as the expected value of $P_{t+1}^c - P_t^c$ exceeds $r_t P_t^c$.

Furthermore, if the specific tax is viewed as a temporary measure, there is an absolute incentive to stockpile because refiner realizations would increase by the amount of the tax when the tax was removed. The specific tax therefore promotes destabilizing inventory behavior.

7. It is acknowledged that this assumes that refiners set prices as if they were operating under a product price control regime. Indeed, that is precisely the effect expected to be observed because with a tax of this form profits can be maximized by building stocks during the initial phases of a disruption and waiting for OPEC to raise prices.

**Table 8-2.** Comparison of the Effect of Three Tax Programs on Changes in Spot Values Under Two Disruptions.

|  | Month After Start of Disruption | | | | | |
|---|---|---|---|---|---|---|
|  | 1 | 2 | 3 | 4 | 5 | 6 |
|  | Dollars per Barrel | | | | | |
| *Small Disruption* | | | | | | |
| $10 Tariff | $29.15 | $28.71 | $27.26 | $26.61 | $25.27 | $23.77 |
| $10 Refiner Tax | 32.15 | 34.65 | 35.57 | 41.01 | 45.11 | 49.84 |
| Windfall Tax | 32.47 | 35.77 | 40.21 | 46.34 | 53.75 | 62.12 |
| *Large Disruption* | | | | | | |
| $40 Tariff | 25.43 | 20.81 | 16.53 | 12.87 | 9.94 | 7.69 |
| $40 Refiner Tax | 37.14 | 47.39 | 59.91 | 73.86 | 88.82 | 104.20 |
| Windfall Tax | 38.29 | 52.20 | 71.51 | 97.49 | 131.81 | 175.88 |

increases in world oil prices. Under the 4 million barrel a day disruption, prices triple with a refiner tax and are increased five-fold with a windfall tax. The large difference occurs because the tax offers the sellers of petroleum products no incentive to get prices up quickly.

From these results it can be concluded that a disruption of a given magnitude can be met effectively by a tariff. A tax on products might also work, but to achieve the same reduction in consumption, the tax would have to be larger than the tariff because the tariff would induce sales from stocks, which would tend to soften the impact of the disruption.

## Variable Disruption Tariffs

The analysis presented above has demonstrated that a tariff on imports of crude and product exerts a greater depressing effect on spot prices than a tax of a similar amount. This means that the effects of a shortage of a specific magnitude may be equally offset through the imposition of a large tax or a smaller tariff.

However, tariff policies have additional powers. A tariff can be used to immediately stop increases in spot prices of oil, or even to defend a particular price. Either result may be achieved by imposing a variable tariff, which would begin at a high level and then be cut to progressively lower values. This section investigates the effectiveness of several variable tariff programs in stopping increases in oil prices started by a disruption or in defending particular price levels during a disruption by using several alternative criteria:

- The percentage of stocks that must be drawn down to achieve a particular target;
- The size of the tariff relative to the predisruption price of oil;
- The cost of delay as measured by the variation in the initial size of the tariff; and
- The combination of tariff size and inventory drawdowns required to defend various price levels.

Essentially, the argument put forward here is that consuming governments can, by the judicious setting of alternative tariff levels, stop or slow the increase in prices on commodity markets and, in turn, prevent escalation of world prices, since producers follow the market. Furthermore, producing countries cannot defeat the adoption of such measures by escalation price of their own because retaliatory increases in the price of oil only serve to further reduce consumption. The decline in consumption creates a surplus that tends to drive prices downward, unless the retaliatory act includes both an increase in prices and a cut in supply.

There are three reasons why such policies can work. First, the imposition of a large initial levy causes a quick cut in consumption. For instance, a $50.00 a barrel tariff is twice as good in cutting consumption as a $25.00 a barrel tariff. However, a large permanent tariff may be excessive and cause a greater reduction in consumption than necessary because of the manner in which demand adjusts to higher prices. Second, consumers would respond to an announcement that prices would fall in the future by delaying purchases. The final reason for imposing a variable tariff is to offer maximum incentives to speculators to sell from their stocks early. Such sales will augment conservation in reducing pressure on the spot market.

*(1) Getting Consumption Down.* The most important reason for introducing a variable tariff of the sort suggested here is to overcome the hysteresis effect in consumer demand. This effect is captured by the difference between the short- and long-term price elasticities of

demand. Numerous studies have shown that both price and income elasticities increase over time. Thus, while a 50 percent increase in price may trigger only a negligible reduction in consumption in the short run, it may induce a fairly large reduction in the long run.

The problem during a disruption is that the long-run response in demand is needed immediately. If the available supply of oil is reduced from 40 million barrels a day to 36 million barrels a day overnight, consuming countries must adopt measures that quickly reduce the demand for oil from 40 million barrels a day to 36 million barrels a day. This can be met by a combination of reduced consumption and stock drawdown. Given the low price elasticities that characterize short run demand, the magnitude of the price increase required to realize an immediate cut of 4 million barrels a day in consumption is on the order of 2030 percent. However, in the long run, a much smaller increase of only 53 percent is required.

In this situation, energy policymakers face a dilemma. They can achieve immediate reduction in consumption by imposing the large tariff, which will quickly lead to the creation of a surplus, or they can impose a smaller tariff that will eventually reduce consumption to equilibrium levels but may allow excess demand during the period of adjustment to push spot prices even higher. This, of course, would eventually push official crude prices higher.

The imposition of a variable tariff addresses this problem. Initially, a tariff is imposed at a level that will shock consumers and cause a maximum reduction in consumption while encouraging sales from stock to balance supply and demand. After the initial response is achieved, the tariff will gradually be lowered as the market aproaches equilibrium.

*(2) Postponement of Consumption.* One reason a variable tariff should be effective in cutting consumption is its effect on consumer expectations. Consumers would presumably postpone discretionary consumption so long as the tariff was imposed in a fashion that guaranteed that prices would decline as the interruption progressed, instead of increasing as they have during past disruptions. The magnitude of the potential postponement in consumption cannot, unfortunately, be quantified because there has been no econometric analysis of the relationship between consumer expectations of future oil prices and consumer purchases. However, it can be noted that demand for oil has, in the past, tended to increase at the outset

of disruptions as consumers have rushed to stock up before price increases occurred.

Evidence from other markets also suggests that expectations as to future prices exert an influence on consumer demand. For instance, automobile manufacturers frequently find that special rebate programs "borrow sales" from future periods rather than increasing total consumption. Furthermore, consumers in the United Kingdom have frequently demonstrated expectational responses during periods just prior to increases in the value added tax. On such occasions it is not unusual to read reports of long lines at stores as consumers attempt to stock up in advance of a price increase. Similarly, when price cuts are proposed, consumers postpone purchases.

*(3) Promoting Sales from Stocks.* The variable tariff would also tend to promote early release of stocks by altering the intertemporal calculations of profits from holding stocks, so that it becomes more profitable to sell early. For example, assume that in the absence of a tariff or tax the price at which a speculator could expect to sell oil would be $30.00 a barrel in the current month but $40.00 a barrel next month due to the price increase resulting from an interruption that is already in progress. Under most circumstances, the logical decision would be to hold on to the oil.

However, suppose that a tariff of $50.00 a barrel is imposed this period and that the government announces its intention to cut the tariff to $35.00 next period. Those holding stocks then would have the opportunity to sell oil this period at $80.00 a barrel or next period for $75.00 a barrel (since the oil already in stock would not be subject to the tariff). Those holding stocks for an additional month would now stand to lose $5.00 a barrel before carrying costs rather than gain $10.00.

The sales from stocks induced by the tariff can be expected to affect oil prices. Thus, if oil prices might be expected to increase from $30.00 a barrel to $40.00 with no tariff, they might increase only to $31.00 or $32.00 with a variable tariff, due to the increased sales of stocks. The knowledge of this simultaneity should act to further accentuate the difference between the moderating effect of a variable rate tariff and a variable rate tax.

In summary, then, a variable rate tariff is a policy tool that can be used to prevent price increases by encouraging inventory drawdown. Furthermore, a variable rate tariff should be expected to encourage a more rapid liquidation than a constant tariff because the latter does not have the same impact on the intertemporal incentive to hold stocks.[8]

## Simulation of the Effect of a Variable Tariff

Simulations of the model of oil market behavior were constructed in order to determine the effectiveness of a variable tariff in stopping an increase in spot market values. All simulations were made under the assumption that the loss in production amounted to 4 million barrels a day and lasted a year.

In constructing the simulations, the basic premise was that consuming countries should follow the example of central bankers who from time to time are called upon to defend currencies against speculative attack. Such attacks may be met by drawdowns of foreign currency reserves, by selling gold, or by swaps with other foreign countries. Each acitivity essentially involves the drawdown of inventories. It is generally acknowledged that such defenses will be less expensive if they are undertaken at the first sign of attack on a currency, provided that the central bankers do not attempt to defend a currency at a level that is unrealistic.[9] For instance, the attempt to maintain a fixed rate of exchange between the dollar and the West German mark in 1971 proved to be impossible due to fundamental changes in the U.S. and West German economies.

Of course, rates of exchange can be defended if the defending country is willing to take actions beyond the use of reserves. Adop-

---

8. The problem is that so long as crude oil prices increase with a fixed tariff, speculators may still profit by holding stocks for an additional period. For instance, a firm may realize a profit $P_t^c + T$ if it sells today, but $P_{t+1}^c + T$ if it holds oil until the next period. The only risk it faces is that the tariff might be removed in period $t+1$. With a variable tariff, the situation is different. The firm must anticipate that the increase in the price of crude will exceed the cut in the tariff for it to be profitable to hold oil. More formally, the rate of change in the tariff, $\Delta T = T_{t+1} - T_t$, must be greater than the expected increase in the price of crude less the cost of carrying the oil. That is,

$$\Delta T \geq (1 - \alpha)(SV_t - P_t^c) - r_t P_t^c - K.$$

9. Keynes was, for instance, highly critical of the British attempt to return to the gold standard at $4.80 to the pound after World War I, arguing that the economic conditions did not support such a high value to the pound. See Kindleberger (1973).

tion of tariffs on imports has, for example, been one mechanism used to reduce balance of payment deficits and defend currencies.

In the case of oil markets, the purpose of policy is to discourage, slow, or stop increases in price, particularly in the spot value of oil, because increases in spot values drive up world prices. As with currencies, the problem is to choose a price that can be defended without exhausting reserves. In the event of a disruption, such a defense can be achieved only by the adoption of an additional measure such as the tariff. Concerning this defensive use of a tariff, the following questions should be answered:

1. How large a tariff must be imposed immediately after a disruption to prevent future increases in spot values of oil?
2. What percentage of consumer country stocks must be used in defense of a given price level?
3. Is a much larger tariff or a much greater use of stocks required if consuming countries attempt to restore prices to their previous level?
4. Is there a danger of retaliatory response from OPEC?

### How Large a Tariff Must Be Imposed in Order to Stop Prices Increases?

The initial simulations assessed the size of the tariff required to stabilize spot values after the initial shock. In constructing the simulation it was assumed that the price of crude oil was $30.00 a barrel prior to the start of the interruption in supply. In the first month of the disruption the simulation indicated that the spot value of crude oil computed on the commodity markets of the world would rise to $37.26 a barrel and that during the first month inventories would be drawn down by 120 million barrels to meet the 4 million barrel a day shortfall.

By trial and error it was determined that a tariff of $31.00 a barrel imposed at the start of the second month would prevent further increases in the spot value of crude oil. This tariff had the effect of increasing consumer prices by 86 percent, an increase which immediately cut consumption by 1 percent. The large tariff also prompted speculators to sell 104 million barrels from inventories.

The tariff was then reduced in the following months by approximately $1.00 a barrel per month, in order to maintain the spot value of crude oil at roughly $37.30. These reductions were required to prevent the development of excess supply from stock sales, which would tend to push down spot values.

The results of this simulation are shown on Table 8-3, where it may be observed that the process of inventory liquidation continued until month 15. By that point, high prices had reduced consumption to the level of supply and stocks began to be rebuilt. The tariff had shrunk to $20.00 a barrel, and consumer prices were only 56 percent above the predisruption level.

The defense described above envisions an attempt by energy policy officials to prevent further increases in the spot value of crude oil. This is not, however, the only defense available. Policy managers could, for instance, attempt to raise prices in a manner that would

**Table 8-3.** Levels of Variable Tariff Required to Meet a 4 Million Barrel a Day Disruption and Prevent Further Increases in Spot Values.

| Month | Loss in Supplies | Tariff Level | Spot Values | Official Crude Prices | Change in Inventories |
|---|---|---|---|---|---|
| | | Dollars per Barrel | | | Million Barrels |
| 0 | 0 | $ 0.00 | $30.00 | $30.00 | 0 |
| 1 | -4 | 0.00 | 37.27 | 30.00 | -120 |
| 2 | -4 | 31.00 | 37.28 | 30.51 | -105 |
| 3 | -4 | 30.00 | 37.39 | 30.98 | - 90 |
| 4 | -4 | 29.50 | 37.42 | 31.43 | - 78 |
| 5 | -4 | 28.75 | 37.44 | 31.85 | - 67 |
| 6 | -4 | 28.00 | 37.42 | 32.24 | - 57 |
| 7 | -4 | 27.25 | 37.35 | 32.60 | - 49 |
| 8 | -4 | 26.25 | 37.28 | 32.94 | - 40 |
| 9 | -4 | 25.00 | 37.34 | 33.24 | - 31 |
| 10 | -4 | 24.25 | 37.36 | 33.53 | - 25 |
| 11 | -4 | 23.50 | 37.33 | 38.80 | - 19 |
| 12 | -4 | 22.75 | 37.24 | 34.04 | - 15 |
| 13 | -4 | 22.00 | 37.25 | 34.30 | - 13 |
| 14 | -4 | 21.25 | 37.30 | 34.61 | - 8 |
| 15 | -4 | 20.50 | 37.26 | 34.89 | - 3 |
| 16 | -4 | 19.75 | 37.24 | 35.11 | + 1 |

restore spot values quickly to the predisruption level. Such a response could be achieved by imposing a much larger initial tariff. Examples of two types of actions are shown on Table 8-4. Here the purpose of the defense is to bring prices back to equilibrium within twelve months (Case $A$) and two months (Case $B$). These policies require higher initial tariffs.

In Case $A$ a tariff of $33.00 a barrel is required initially, while in Case $B$ a tariff of $51.00 is required. However, because the initial tariff achieves a greater cut in consumption, the size of the tariff is lower after the second month. This suggests that some adverse macro economic effects might be avoided by the imposition of a very large initial tariff.

### What Percentage of Consumer Stocks Must Be Used to Defend a Particular Price?

The percentage of consumer stocks required to defend a particular price level depends first upon the price increase allowed by policy and the rapidity with which the tariff is imposed to bring prices in line with the target price. In general, a higher percentage of inventories must be used to stabilize the world price at a predisruption level than to stabilize them at a higher level. However, the differences are not all that great. For instance, the stock drawdown simulations shown on Table 8-4 where prices are forced back to $30.00 a barrel are 0.5 percent greater than the stock usage in Table 8-3 where the price is allowed to stabilize at $37.00 a barrel.

Stock drawdown is increased significantly, however, if the crisis is allowed to persist and if the interruption is not met by some other management technique, such as rationing by queuing. For instance, if a 4 million barrel a day interruption is allowed to last for a year without policy response, then a stock drawdown of 1,440 million barrels would have to occur. On the other hand, with a tariff, drawdown would only have to amount to 700 million barrels.

### Is a Larger Tariff Required to Restore Prices to the Predisruption Level?

The simulations offered in Table 8-4 suggest that the size of the tariff required to restore prices to the predisruption level does not have

**Table 8-4.** Levels of Tariffs Used to Restore Spot Values to Predisruption Levels.

| Month | Loss in Supply | Case A: Slow Adjustment in Spot Values ||| | Case B: Quick Adjustment in Spot Values ||||
|---|---|---|---|---|---|---|---|---|---|
| | Million Barrels a Day | Tariff Levels | Spot Crude Values | Official Crude Prices | Change in Inventory | Tariff Levels | Spot Crude Values | Official Crude Prices | Change in Inventories |
| | | Dollars per Barrel ||| Million Barrels | Dollars per Barrel ||| Million Barrels |
| 0  |  0 | $ 0.00 | $30.00 | $30.00 |    0 | $ 0.00 | $30.00 | $30.00 |    0 |
| 1  | -4 |   0.00 |  37.27 |  30.00 | -120 |   0.00 |  37.27 |  30.00 | -120 |
| 2  | -4 |  33.00 |  36.52 |  30.51 | -105 |  51.00 |  30.08 |  30.51 |  -99 |
| 3  | -4 |  32.00 |  35.57 |  30.93 |  -90 |  22.00 |  30.22 |  30.48 |  -86 |
| 4  | -4 |  30.00 |  34.75 |  31.25 |  -78 |  23.00 |  30.17 |  30.46 |  -80 |
| 5  | -4 |  29.00 |  33.78 |  31.50 |  -66 |  23.00 |  30.15 |  30.44 |  -70 |
| 6  | -4 |  26.00 |  33.32 |  31.66 |  -57 |  23.00 |  30.16 |  30.42 |  -62 |
| 7  | -4 |  26.00 |  32.55 |  31.78 |  -48 |  23.00 |  30.17 |  30.40 |  -54 |
| 8  | -4 |  24.00 |  32.05 |  31.83 |  -41 |  23.00 |  30.18 |  30.39 |  -48 |
| 9  | -4 |  24.00 |  31.28 |  31.85 |  -34 |  23.00 |  30.15 |  30.37 |  -42 |
| 10 | -4 |  22.00 |  30.80 |  31.81 |  -29 |  23.00 |  30.07 |  30.36 |  -37 |
| 11 | -4 |  21.00 |  30.43 |  31.74 |  -24 |  23.00 |  29.92 |  30.34 |  -33 |
| 12 | -4 |  21.00 |  29.88 |  31.64 |  -20 |  23.00 |  29.67 |  30.31 |  -30 |

to be much larger than the tariff required to prevent prices from rising further. In fact, after the first few months of a disruption, the level of the tariff is irrelevant, as long as prices have been stabilized, because the world price of crude oil will pick up whatever consuming countries fail to capture through the tariff. Thus, the consumer will pay the same price, whether it is achieved by a sudden increase in price achieved through the imposition of a tariff, or through the imposition of higher world oil prices. Since the latter involves transfers of wealth that have macroeconomic impacts while tariffs do not appear to have macroeconomic impacts if they are recycled, it is suggested that the large tariff may be better.[10]

### Is There a Danger of Response by OPEC?

One of the arguments against the use of a tariff has always been that it would trigger an increase in world oil prices. After all, the argument goes, if consumers want higher oil prices, the members of the cartel will happily oblige. This study rejects this view because it finds that OPEC tends to follow the spot market. Thus, it has been suggested that tariffs imposed to stabilize increases in spot market prices would, perforce, tend to prevent world oil prices from increasing.

However, it is possible that petroleum exporters would ignore the logic suggested above and attempt to raise prices. If such an action were taken, it obviously would have to be accompanied by further reductions in the production of oil. For instance, rough analysis of the model of oil market behavior suggests that producers would have to reduce output by an additional 4 million barrels a day beyond the initial 4 million barrel a day cut if they were to succeed in pushing prices up to, say, $50.00 a barrel. Such a response could, of course, be countered by further tariff increases in consuming countries, the result of which would be to heap hardship upon citizens in both consuming and producing states.

This type of behavior may be simulated in the model of oil market behavior. However, since the number of possible variations is limitless, no results are presented. Instead it is noted that:

- Retaliation by OPEC in price alone increases the conservation effect of the variable tariff and thus increases the surplus, driving

10. Calculations by the U.S. Congressional Budget Office (1978) indicate that a fully recycled tariff has no impact on real GNP or employment.

spot prices down. With a $30.00 a barrel tariff and an artificially inflated OPEC price of $50.00, it is possible to quickly drive spot values below $10.00.

- The decline in spot values can be prevented if OPEC combines a price increase with a second-round cut in production.

The problem of retaliation can be circumvented by imposing a large tariff for a very short period of time and then scaling it back. Table 8-4 shows that a $51.00 a barrel tariff imposed for one month followed by a tariff of $23.00 for the next eleven months was adequate to balance supply and demand. It is presumed that the imposition of a one-month tariff would have fewer undesirable effects than a permanent tariff. Furthermore, the $23.00 tariff could be adjusted downward as official OPEC prices were increased.

### Economic Consequences of the Imposition of a Tariff

The economic consequences of imposition of a disruption tariff or tax could be severe. In addition to the loss in output and income caused by the increase in oil prices, the federal government's bite in revenues from tariffs and the existing windfall profits tax would be increased. The fiscal drag caused by these increased taxes would tend to further reduce output and income unless they were quickly recycled.

Most studies suggest that if the tax revenues were recycled quickly, so that there were no net increase in federal revenues, the net loss to the economy could be minimized. These studies indicate that the recycling process would have to be quick and that the Federal Reserve would have to accommodate the higher price level by increasing the rate of growth of the money supply. The evidence presented below suggests that the money could be recycled quickly in the United States under certain very restrictive cases. European countries should have less trouble because they could alter value added taxes when they imposed the tariff.

## Would a Tariff Increase or Decrease the Pain?

It is generally acknowledged that a disruption in oil markets would adversely affect real outputs and incomes. This effect has been clearly identified by Fried and Schultz (1975:4), who describe the phenomena surrounding the 1974 debacle as follows:

> In essence, the initial impact of the oil price increase can be compared to the imposition by the producers of oil of a large excise tax, the proceeds of which were not immediately used to buy goods or services. Consumers in the importing nations paid more for energy and therefore had less to spend for other products.
>
> As a consequence, sales, output, and employment were reduced in the consumer-goods industries. Total demand and output fell, not only by the initial amount of loss in consumer demand, but still further through the typical cyclical process in which the initial reductions in employment and income were the cause of still further declines in demand, in output and in jobs. Moreover, the loss of output and income in each country provides an export market for the others.

More recently, the U.S. Congressional Budget Office (1978), the U.S. Department of Energy, and several private organizations have all issued studies that trace out the effect of higher oil prices on the U.S. economy. All reach the same conclusion.

None of these studies has attempted, however, to address the issue of the effect of the imposition of a tariff at the time of a disruption. Plummer (1981:29) has addressed this issue and argues, "Disruption tariffs require a very high level of micro economic and macro economic sacrifice during disruption years, just when the economy may be least able to bear it. In contrast to the stockpile fill rate that acts as an automatic stabilizer, a disruption tariff acts as an automatic destabilizer. The economic losses that result from an oil supply disruption are the result of the economy's inability to adjust fast enough to rapid oil price increases. A disruption tariff increases domestic oil prices even above the level that the disruption itself causes and thus hits the economy at its most tender point." Plummer reaches this conclusion by studying the results of simulations of several large-scale models.

If one assumes that the level of world oil prices is a given and that the imposition of a tariff or tax will not have an impact on the behavior of oil inventories, then this view is absolutely correct. However, the imposition of a tariff can moderate the increase in oil prices and prevent them from overshooting a new equilibrium level and can induce private speculators to sell their stocks, further dampening the increase in oil prices. These observations lead to the conclusion that the Plummer view is correct in a stylized, abstract world of large-scale economic models but irrelevant to the real world.

Regarding the prospects and problems faced by OECD economies in the face of a disruption, this study accepts the view first put forward by Fried and Schultz (1975) and demonstrated many times since, that any policy that successfully discourages increases in oil prices while simultaneously recycling any increased tax and tariff revenues will have positive economic benefits. The reader is referred to the U.S. Congressional Budget Office (1978) study of the economic effects of a recycled tariff, which shows that there would be no macroeconomic effects from a fee.

### Recycling Tariff Revenues

The conclusion that the imposition of a large tariff would tend to moderate the impact of a disruption by holding down prices assumes that the federal government can quickly recycle the increased revenue from the tariff. This assumption has been validated in other studies (U.S. Congressional Budget Office 1978; Verleger and Sheehan 1974). In each study, economic models of the U.S. economy have been subjected to price shocks, increased taxes, and increased federal disbursements that effectively returned purchasing power to consumers instantaneously. In general, it is easy to perform such simulations with econometric models but far more difficult to achieve the same effect in the real world. Legislative authority is required to impose a tariff and redistribute the proceeds. Furthermore, it takes additional time to redistribute the proceeds of a tariff once the authority has been legislated.

Under Section 232(b) of the Trade Expansion Act of 1962, the president of the United States currently has authorization to impose an emergency tariff. That legislation authorizes the president to adjust imports of crude oil and products if it is determined that the volume of imports is such as to impair national security. For the tariff approach to succeed in reducing oil imports without causing great harm to the economy would require legislation comparable to Section 232(b) authorizing distribution of the proceeds of the tariff. Such legislation presently does not exist. If a tariff were imposed, the President would have to ask Congress to enact legislation authorizing a tax cut or some other mechanism to recycle the revenues. As Ronald Reagan, Jimmy Carter, and every other U.S. president have discovered, the legislative process takes time.

The problems do not end, however, when action on tax legislation is completed. Additional time is required to change regulations and write checks. In the case of a tariff on oil, most economists would council that Congress write a very simple tax bill that offsets the increased tariff receipts by temporarily increasing Social Security payments and reducing personal tax liabilities and withholding rates. In terms of tax regulations, such changes would be by far the easiest to implement. However, as Emil Sunley (1980), former Deputy Assistant Secretary of the Treasury for tax policy, noted in testimony before Senator Bradley, even implementation of these changes would be difficult.

Sunley identified three major problems. First, in drafting legislation there would be great concern that every individual receive one and only one rebate. Second, since much of the money would be returned by reducing personal tax withholding, those doing the withholding (firms, governments, private individuals) would want at least six weeks to implement the changes. Third, the tax liability of some low income individuals would be so small that they would have to get back more than was being withheld.

The drafting problem raised by Sunley concerns the problem of designing a tax under which every individual receives one rebate and no individual receives two. As Sunley (1980:82) noted in his testimony, "Then you have to work out the problems of some individuals who might not be reached under any of your mechanisms, and other individuals who might be reached more than once. . . . The problem that we always run into in this area and the problem when we looked at something as relatively simple as the $50.00 rebate in 1977, is that we wanted to give a reduction to everyone, but only one rebate to each person."

One way to get around the one man/one rebate problem is to refund the proceeds of the tariff by a general tax reduction. However, as Sunley notes, some individuals pay no federal income taxes. Of course, a portion of these individuals could be reached by increasing Social Security benefits, but an increase in Social Security payments combined with a general tax reduction would confer double benefits on some individuals (Sunley 1980).

Sunley also addressed the problem of implementing reductions in withholding, stating that at least sixty days would be required to implement the cut. He (Sunley 1980:82) explained, "It usually takes about 15 days just to get the withholding forms printed and ready for distribution. The corporations and businesses generally want 30 to 45 days to make a change in their withholding, just to do the reprogramming of their computers. This is particularly true of larger companies which may have the payroll function decentralized and small businesses that manually use the withholding tables."

Finally, Sunley addressed the problem raised by the fact that the tax liability of some low income individuals would be less than the size of the rebate authorized by Congress. In these cases mere adjustment of withholding rates would not serve the purpose of fully recycling the tariff. One alternative suggested by Sunley (1980:84) would have the employer advance the employee the full amount of the rebate legislated by Congress and then have the federal government reimburse the employer: "So it does seem to us that you might have to develop a system which would get federal moneys to the employers to pay the rebate. The system you would have to develop would be the opposite of the federal deposit system now. We have a system where the employers are required to deposit moneys with certain banks. We would have to develop the opposite of that, where the federal government, in effect, transfers moneys to the banks a couple of days before payday, and then the businesses could actually withdraw the money to make the rebate payments."

In summary, the problems involved in the recycling of tax revenues received under any tariff or tax in the United States are substantial. The solution is not as simple as changing the inputs to an econometric model.

One principal advantage enjoyed by other countries is the use of the value added tax (VAT) rather than income taxes. Because they are basically sales taxes, these taxes can be adjusted quickly. The short-term economic effect of reducing value-added taxes would be the same as that of reducing income tax rates simultaneously with imposing a tariff. Each would hold purchasing power constant. In the long run, however, changes in the VAT and the personal income tax might have slightly different effects because a reduction in the VAT might reduce the incentive to save more than a change in the personal income tax.

## THE PROBLEM OF RETALIATION

The imposition of a tariff on imported oil by the United States, or by all members of the OECD acting in concert, creates the risk of confrontation. The imposition of a tariff may cause the members of OPEC to retaliate by increasing their prices. The risks of such retaliation have, in the past, caused various administrations to drop the idea of imposing a tariff.[11] This same view is espoused by Plummer (1981:29), who argues that the imposition of a tariff might not only prompt a price response by members of OPEC but might also cause reductions in supply.

> Another problem with a disruption tariff is the uncertainty with regard to possible Saudi retaliation. The Saudis may be less likely to retaliate to stockpiling than to a regular tariff (in nondisruption years), because a tariff is more closely related to price. The Saudis are also more likely to retaliate to a disruption tariff than to a regular tariff because it is much easier for them to retaliate during a disruption period.

The portion of the Plummer view that is relevant here is his concern about the response of Saudi Arabia to the imposition of a tariff during a disruption. The questions to be answered are:

1. Why would the Saudis want to respond?
2. Would it matter?
3. Would other members of OPEC also respond? and
4. Could the developed nations also respond again?

### Why Would Saudi Arabia Respond?

Adelman (1980:51) has noted, "The Saudis, their neighbors and others are fine tuning a cartel with coarse instruments." Further-

---

11. For instance, in the summer of 1978 Secretary of Energy James Schlesinger suggested that a tariff on oil might be imposed if Congress did not get on with the job of passing the National Energy Plan. His public statements were greeted with substantial disgust by members of OPEC.

more, as some have noted, the optimal long-run price of oil for the Saudis is probably lower than the present level. These statements suggest that Saudi Arabia should currently be pushing for lower oil prices and would, in a disruption, support any effort that might prevent oil prices from surging to excessive levels. The actions of Saudi Arabia during the spring of 1981 when they "glutted the world oil market" tend to support such a view.

If this view is correct, and if Saudi policy holds to a position that would try to avoid sudden large increases in world oil prices from the $30.00 a barrel level, then it must be assumed that Saudi Arabia would support the imposition of *temporary* tariffs during a disruption so long as it was clearly understood that the tariffs were imposed only to calm the spot market and would continue only until the disruption persisted.

### Would the Views of Saudi Arabia Matter?

A second problem with the Plummer position is his presumption that the views of Saudi Arabia would matter. It is certainly possible that the particular disruption might involve a loss in supply from Saudi Arabia. In this case, the views of the Saudis, like those of the former Shah in 1980, would become irrelevant.

Also, the loss of production from Saudi Arabia would probably represent the greatest threat to the world economy. The other members of OPEC are endowed with smaller reserve bases and thus are less interested in maintaining an oil market into the twenty-first century. This means they would be more inclined to raise prices while maximizing output in order to take the fullest possible advantage of the situation. The use of a variable tariff approach would then be more important in the case of a loss of Saudi production than in the case of any other type of disruption.

On the other hand, if the loss of supply did not affect Saudi Arabia, the Saudi view would matter because they could counter any move made by consuming countries. However, as noted in the previous section, the Saudis would presumably welcome a temporary action that prevented prices from increasing excessively.

### Would Other OPEC Nations Respond?

It is quite possible that other OPEC nations might respond to the imposition of a tariff by raising their own prices or by cutting supply. An increase in prices would have the same effect as the imposition of a tariff. It would speed up the process of adjustment. Thus, retaliation by OPEC in response to the imposition of a tariff would bring oil prices quickly to the levels they would have reached in any event. Therefore the question of OPEC response is moot.

### Could Consumer Countries Respond Again?

There is no reason why consumer countries would have to be mute in response to an attack by OPEC on their imposition of a tariff. Indeed, an OPEC response would offer them two choices: They could leave their tariff in effect and thus threaten to create a glut that would force petroleum exporters to reduce supply or lower price, or they could respond to the increase in the price of oil by reducing the tariff. If the tariff is imposed only for the purposes of accelerating the process of adjustment, then the second action would be best.

The use of a quota/auction throughout the OECD instead of a tariff would be one means of avoiding retaliation. The imposition of an OECD-wide quota together with an auction would leave the setting of world crude oil prices to OPEC, with the value of auction tickets representing the difference between the prevailing world crude oil price and the equilibrium market price.

Such a policy would be the least confrontational approach to adopt toward petroleum exporters during a disruption because it would declare that consuming countries recognized a shortage and had established a market mechanism to bring consumption into line. However, there are certain potential drawbacks to this proposal. It would have to be imposed throughout the OECD so that there would be no accusations that one country was absorbing more of its share of the shortage than another. In addition, serious problems might be encountered in distributing the receipts among countries. Finally, the initial price of quota tickets might be very high due to uncertainty as to future supply, and this initial high price could give

OPEC the wrong signal.[12] Although a quota/auction could offer an attractive and less confrontational alternative to the tariff, it also appears that more research is required—research primarily directed at the behavior of the value of quota tickets in the early phases of a disruption.

## Is Retaliation a Problem?

Retaliation by petroleum exporters to the imposition of a tariff at the time of a disruption should not be viewed as a problem. The tariff should not be viewed as a weapon, nor the disruption as a war. Instead, the disruption should be viewed as a form of cancer, and the tariff as a form of chemotherapy; higher OPEC prices are another, less satisfactory, form of therapy.

## CONCLUSION

This chapter has shown that tax measures offer one mechanism for ameliorating the adjustment problems experienced during a disruption and that all taxes are not equally effective. The quota/auction and tariffs on the importation of oil appear to be most effective because they simultaneously induce a quick response by consumers and cause speculative stocks to be liquidated, calming the disruption. Product taxes, or refiner excise taxes, were shown to be less effective because they affect only consumer demand. Finally, windfall profits taxes were determined to be of no use at all because they failed to encourage speculative sales from stocks or quickly lift consumer prices.

The preferred option described here would be a tariff imposed initially at a very high level and then scaled back over time. This scheme would encourage maximum immediate conservation and early speculative liquidation of stocks. Both actions would tend to dampen increases in world oil prices. It was also noted that the macroeconomic impacts of the imposition of a tariff could be severe

---

12. The price of quota tickets should decline over time as demand adjusts to higher prices and as stocks are sold. Thus, the price of quota tickets should roughly follow the graduated tariff.

but more desirable than the economic impact that would result from letting the price of oil escalate in an uncontrolled fashion.

The imposition of a tariff in the United States does, however, pose one very messy bureaucratic problem because no easy recycling mechanism is available. Furthermore, the time required to make the necessary adjustments in the tax code, withholding tables, and so forth, could be substantial.

Finally, the tariff should not trigger retaliation from OPEC, but in any case, retaliation by OPEC in the form of higher prices is a response that should not prevent consuming countries from imposing a tariff. A better response would be to lower the level of tariffs since the increase in OPEC prices also would speed the adjustment to the new level of supply—the adjustment that is the raison d'etre of the tariff.

# 9 POLICIES TO ENCOURAGE BUILDING INVENTORIES BEFORE DISRUPTIONS

> We must encourage rather than discourage speculation on the holding of oil stocks throughout our economy. We must encourage people to buy cheap and sell dear, for such behavior is price stabilizing and in the economic interests of the nation. I know that such a principle will lead some to gag, but it is a simple economic fact that by encouraging speculative behavior we will build up stocks for profitable release—but for release—when oil shocks and high prices occur.
>
> William Nordhaus in Hearings Before the Subcommittee on Taxation and Debt Management of the Senate Committee on Finance, *Special Oil Taxes* (1980:32).

Throughout this discussion it has been emphasized that inventories play an important role in the process of moderating disruptions: They are the standby source of supply. The building of speculative or precautionary inventories, however, is an expensive and probably money-losing, proposition. Thus, the governments of consuming countries have had a difficult time convincing the public to build stocks. In some countries, such as Germany, Japan, and France, government fiat has replaced persuasion. In other countries, such as the United States, the government has not elected to impose such regulations, although it has the authority but instead has chosen to build federally owned stockpiles.

Several reasons may be suggested for the reluctance of the private sector to develop standby stockpiles, including the expense, the low probability of profit, the high probability that speculators will be denied their profit through the imposition of governmental controls, and the prospect of available, if limited, low-cost supplies during a disruption. This chapter examines several programs for increasing private sector willingness to hold stockpiles.

## INCENTIVES TO BUILD SUPPLIER INVENTORIES

High storage costs, low profits, and the threat of imposition of price controls or taxes constitute significant barriers to the development of stocks by both suppliers and consumers. The high costs are primarily a consequence of the high price of oil and high interest rates. At interest rates of 15 to 20 percent and with oil prices above $30 a barrel, the annual interest cost of holding a barrel of oil will range from $6.00 a barrel to $8.00 a barrel, exclusive of any direct storage costs. This means that a firm adding a barrel of oil to its inventories for a year must forego income of at least $6.00 to $8.00 per barrel of stocks, assuming that the purchaser has a good credit rating. Under normal circumstances, a prudent investor would not invest in an additional speculative barrel of inventories unless he expected the sale price of the oil to be $8.00 higher than the current price. Indeed, there is a general and obvious rule for commodity trading (the arbitrage rule) that dictates that stocks be acquired only when the expected price, $E(P_{t+1})$, exceeds the price in period $t$ by more than the risk adjusted interest cost plus storage cost of holding the commodity from the current period to the next period. This implies that stocks should be acquired if

$$E(P_{t+1}) > (1 + r_t)P_t , \qquad (9.1)$$

while they should be sold if

$$E(P_{t+1}) < (1 + r_t)P_t , \qquad (9.2)$$

assuming marginal storage costs are zero.

This basic rule can be applied to oil markets during the 1973–81 period by comparing the year-to-year change in the price of oil with

Table 9-1. Comparison of Year-to-Year Changes in Official Sales Prices of African Light Crude (*January prices*).

| Year | Price at Which Crude Could Be Purchased in Year t-1 ($P_{t-1}$) | Cost of Crude Purchased in Year t-1 Adjusted by Inventory Cost[a] | Price at Which Crude Could Be Purchased in Year t | Arbritrage Signal in t-1 if Price in Year t Is Known |
|------|------|------|------|------|
| | | *Dollars per Barrel* | | |
| 1973 | $ 2.80 | $ 2.95 | $ 3.10 | Buy |
| 1974 | 3.10 | 3.35 | 10.75 | Buy |
| 1975 | 10.75 | 11.91 | 11.80 | Sell |
| 1976 | 11.80 | 12.72 | 12.84 | Buy |
| 1977 | 12.84 | 13.67 | 14.33 | Sell |
| 1978 | 14.33 | 15.30 | 14.33 | Sell |
| 1979 | 14.33 | 15.62 | 14.84 | Sell |
| 1980 | 14.84 | 16.72 | 34.67 | Buy |
| 1981 | 34.67 | 39.94 | 41.00 | Buy |

Source: Price data from PIW. Costs calculated by the author.
a. First column is adjusted by the prime rate of interest in the year of purchase.

the interest cost of holding the crude (assuming that $E(P_{t+1}) = P_{t+1}$). In particular, if the change in the cost of acquiring a barrel of African light crude oil (the most market sensitive crude)[1] at the beginning of the year is compared with the price the crude would have brought at the beginning of the next year, it is clear that it would have been profitable to speculate in five of the nine years between 1972 and 1980 (1972, 1973, and 1975 and then again in 1979 and 1980. See Table 9-1.). On the other hand, the profitable strategy in 1974, 1975, 1977, and 1978 would have been to sell. (Note that 1976, 1977, and 1978 were years when there was a generally acknowledged surplus.)

A slightly different conclusion is reached when changes in crude prices from the end of the year to the end of the year are examined. However, based upon the behavior of the market, we conclude that

---

1. It was noted in Chapter 2 that the African producers are generally the most responsive to changes in the market, raising and lowering the price of oil quickly in response to changes in spot values.

the profitable rule for speculation during the 1970s would have been to buy at times of shortage and sell during times of glut.

Of course, the calculations shown on Table 9-1 are limited because they assume that stocks are only held for a one-year period, and inventories acquired for the purpose of meeting a disruption would presumably be held for several years. Furthermore, the calculations make no adjustment for risk, an obvious oversight. A more appropriate test is to determine the break-even price if oil were acquired at year $t$ and sold after $n$ years. For example, suppose a speculator had acquired oil in early 1976 and held it until the beginning of the Iranian crisis in January 1979. The acquisition cost of the oil would have been $12.84. To break even three years later, the speculator would have had to sell the oil for $15.88 a barrel,[2] but the price of oil in January 1979 was $14.84 or less. Thus, the speculator stood to lose at least a dollar by performing this public service. On the other hand, if the temptation to sell at the start of the Iranian disruption was resisted and the oil remained in stocks for another year, the investment would have been quite profitable but would have served no public good.

The same situation would have applied to oil acquired at other times during the mid-1970s. In general, the prices of oil prevailing during the early phases of disruptions were less than the break-even prices required to make speculation an appealing option.

The calculations of these break-even prices are shown on Table 9-2, which also suggests that the speculator who did not sell oil from inventories in early 1979 but elected to hold them for an additional year would have made a handsome profit. However, the profit would have been even greater had he waited to acquire the oil in 1979 rather than tying up his funds for two or three years.

Thus, the clear conclusion to be drawn from this analysis is that the development of speculative inventories during times of surplus would not have been a profitable proposition in the 1970s. This conclusion holds even if no regulations had been imposed. Indeed, the calculations shown here assume that the holders of inventories were free to sell their oil at prevailing free market prices.

This suggests, then, that the development of precautionary or speculative stocks would require some type of public policy action

---

2. We estimate the $15.98 by multiplying the $12.84 by the interest cost of borrowing money for 1976, 1977, and 1978, i.e.,

$$\$15.98 = 12.84 \times (1 + r_{76})(1 + r_{77})(1 + r_{78}).$$

**Table 9-2.** Calculation of Break-Even Prices for Crude Oil Held More than One Year *(Based on African Light)*.

| Year Oil Acquired | Acquisition Price | Break-Even Price If Sold In: | | | | | |
|---|---|---|---|---|---|---|---|
| | | 1976 | 1977 | 1978 | 1979 | 1980 | 1981 |
| | | *Dollars per Barrel* | | | | | |
| 1975 | $11.80 | $12.72 | $13.60 | $14.73 | $15.84 | $17.85 | $20.58 |
| 1976 | 12.84 | | 13.67 | 14.66 | 15.98 | 18.01 | 20.76 |
| 1977 | 14.33 | | | 15.31 | 16.70 | 18.81 | 21.68 |
| 1978 | 14.33 | | | | 15.63 | 17.61 | 20.30 |
| 1979 | 14.84 | | | | | 16.72 | 18.73 |
| Actual Price | | 12.84 | 14.33 | 14.33 | 14.84 | 34.67 | 41.00 |

Note: Calculations no not include any adjustment for the cost of storage.

that would either reduce the cost of holding stocks or increase the expected reward offered to those who develop stocks during periods of surplus.

One option for increasing returns suggested by Lane (1980) and Mampe (1981) is to exempt from U.S. income tax liabilities profits made by the sale of speculative inventories during a disruption. Under this scheme, a firm or individual that built speculative stocks during a period of surplus and then sold them during a disruption would pay no income tax, or a smaller tax, on the profit realized from the sale of those stocks. Such a policy would approximately double any gain realized by selling speculative stocks and thus cut in half the break-even prices.[3]

An alternative proposal would be to subsidize the cost of storage by reducing the interest rate paid for funds borrowed to purchase oil stocks. Such a scheme would effectively reduce break-even prices by lowering the annual cost of carrying the oil. This could be done either by direct loans or through tax subsidies. If provided by the direct approach, the government would subsidize the interest rates paid for loans to acquire additional oil stocks in the same way that it subsidizes home and farm loans. Alternatively, the subsidy could be paid through the tax system by providing increased deductions or tax credits for interest costs incurred in connection with storage of the oil.[4]

---

3. This can be shown by noting that the gain from the sale of stocks is equal to the sales prices of the asset minus the present value of the asset:

$$\text{Gain} = P_t - P_0 \prod_{i=1}^{t} (1 + r_t) .$$

Ordinarily, the tax liability on the gain would be given by

$$\text{Tax} = \tau (P_t - P_0) ,$$

where $\tau$ is the tax rate, because interest is deducted from prior years' income taxes. However, for speculative stocks which were exempted from taxation, the gain would be given as

$$P_t - P_0 .$$

This would mean that the profit on speculative stocks was $1/(1 - \tau)$ times as good as the gain on ordinary stocks.

4. Interest payments are ordinarily deductible from taxable income as an expense of doing business. Thus, the interest, $r_t P_0$, could be deducted from income in year $t$, for a barrel of oil purchased in year $o$, where $r_t$ represents the interest rate in year $t$ and where $P_0$ represents the acquisition cost in year $o$. This payment is worth $\tau r_t P_0$ in after-tax dollars to the firm paying a marginal rate $\tau$.

A tax credit would permit the firm to deduct the full amount of interest payments from the income tax liability. Thus, the increased after-tax benefit would be worth an amount

The following illustration may clarify the way in which the two incentives would work. Suppose Smith Oil acquires 1 million barrels of oil to hold as speculative stocks on January 1, 1981, at a price of $30.00 a barrel and borrows the $30 million to pay for the oil at an interest rate of 20 percent. Under present tax law, Smith Oil would then be entitled to deduct interest payments of $5 million from its 1981 tax returns, assuming it held the oil in inventories all year. If in 1981 Smith Oil's pretax profits would have been $500 million dollars and tax liability would have been $260 million dollars, then the payment of interest on the speculative inventories would reduce the profit by $5 million to $495 million and reduce tax liabilities to $257.4 million.

**Effect on Smith Oil of Purchase of 1 Million Barrels of Stocks.**

|  | Pretax Profits | After-Tax Profits |
|---|---|---|
| Income before investing in oil | $500 million | $260 million |
| Impact on profits of $30 million loan at 20 percent | −5 million | −2.6 million |
| Net position after investment | 495 million | 257.4 million |

The same type of calculation would be repeated in succeeding years until the oil was sold, when the firm would realize a gain in income due to the rise in the price of oil. For instance, if the oil were sold in 1984 at a price of $50 a barrel, the firm would realize a gain of $20 million. Under existing law, this gain would be taxed as ordinary income. Thus, if other 1984 income were $600 million and the tax rate were still 48 percent, the effect of the sale of stocks on the firm's financial position would be:

**1984 Income Statement of Smith Oil, Inc.**

|  | Pretax Profits | After-Tax Profits |
|---|---|---|
| Income before sale of inventories* | $600 million | $312 million |
| Profit from sale of stocks | 20 million | 10.4 million |
| Net effect of investment | 620 million | 322.4 million |

*Assumes stocks sold on January 1, 1984, so that no interest liability was incurred.

---

$(1 - \tau) r_t P_o$. The effect of the tax credit would be to reduce the interest cost by $\tau$ when speculative projects were compared with ordinary storage. Thus, if project $A$ entitles the speculator to take a tax credit on interest payments but project $B$ does not, the present discounted values of $A$ and $B$ would be computed using an interest rate of $\tau r_t$ for $A$ and $r_t$ for $B$.

The proposal to exempt the gain on the sale of the inventories from taxation as ordinary income would alter the 1984 after-tax income statement of Smith Oil. After-tax profits would be increased from $322.8 million to $332 million, almost doubling the gain on the sale.

The proposal to subsidize interest payments on speculative stocks would not affect the 1984 income statement (because the oil was assumed to be sold at the first of the year). However, the income statements in the preceding years would be changed. For instance, in 1981 after-tax income would be increased from $257.4 million to $260 million as a result of payments of interest subsidies to the firm. Thus, the subsidy payments could leave the firm's after-tax income uneffected.

In principal, these two types of proposals (exemption of the gain from taxation or subsidization of interest payments) could both induce firms to hold larger stocks. However, the interest rate subsidy would probably induce greater stock holding since investments must promise an early return to be profitable when interest rates are high. The payment of the benefit annually would probably have a greater present discounted value because the value of a financial gain at some uncertain date in the future will be small even if the gain is known to be large. Also, the reward promised by the exemption from taxation of sales of speculative stocks during disruptions cannot be offered with assurance, while the subsidy on interest can. This difference occurs because the subsidy on interest expenses is paid annually while the stocks are held and thus cannot be changed by future legislation. On the other hand, the exemption of speculative gains from taxation requires both the passage of legislation by Congress at the start of the program and agreement by future Congresses not to alter the legislation.[5]

Both of these proposals have two problems. First, they would be difficult to administer because the incentives presumably would

---

5. This problem, which is met many times in the remaining chapters of the book, occurs because it is not constitutionally possible for one Congress to commit a future Congress either to take or not to take some action. Thus, a tax law could be changed to favor speculators during a period of surplus, and then changed again at the start of a shortage to deny the anticipated benefit to speculators. Naturally, speculators recognize this fact and may thus tend to be highly skeptical of promises of rewards in the future. The same reasoning applies to promises not to impose price controls. One Congress and one chief executive may promise not to impose controls, and then the next chief executive or Congress may withdraw that promise.

apply only to speculative stocks. Stocks held in the absence of an incentive would not qualify for an interest rate subsidy, and sales of inventories that normally would take place due to seasonal fluctuations in demand would not qualify for exemption from the profit tax.[6] This means that agreements and regulations defining normal and speculative stocks would be required, giving rise to a new bureaucracy.

A second and more troubling disadvantage is that these proposals offer no incentive to sell the stocks at the time of a disruption. They do not even reduce the incentive to acquire stocks during a disruption. Thus, it is noted now and will be argued below that a program designed to encourage stock building will be of no use unless it is accompanied by a program that promotes the sale of stocks at the start of a disruption.

## INCREASING THE POTENTIAL PROFIT FROM HOLDING STOCKS

It would also be possible to affect the level of inventories by altering future prices. The United States has tried this method during the last eight years, but the incentives had the perverse effect of discouraging storage instead of encouraging it. The incentive was, of course, price controls.

Price controls slow the rate of increase in prices. This tends to make the building of speculative inventories less financially attractive. There appears to have been no incentive to build stocks under the controls imposed in the United States between 1973 and 1981. In particular, a firm acquiring crude oil in year $t$ for the purpose of speculation was required to use the price paid for the oil in determining the costs of its products when the oil was refined, regardless of the number of years the oil was held. Thus, a refiner that purchased a barrel of imported oil in 1975 for the purpose of building inventories was required to use the 1975 cost in computing selling prices, even if the oil was taken from inventory in 1981. This requirement obviously made it idiotic to build stocks.

---

6. The economic justification for such a policy is cost minimization. Presumably, tax agencies or energy officials would not want to sacrifice federal receipts or make interest subsidy payments for stocks that would have been held anyway.

Mandatory allocations represent another type of program that discourages firms from developing speculative inventories. These were established in the United States from 1974 to 1981 at two levels: crude supply and product distribution. Under the crude oil allocation program, certain large refiners were periodically required to supply specific volumes of crude oil to a second group of smaller refiners at prices based on the selling firm's average cost of imports. Product allocation programs established a set of priorities that all companies were required to follow in distributing product.

The economic effect of a crude oil allocation program is to reduce the incentive to hold stocks by those firms entitled to receive allocations, since the price they expect to pay for crude oil in the event of a disruption is reduced. Indeed, in the past, one of the justifications expressly stated by the Department of Energy in granting allocations to smaller firms was to preserve the competitive position of such firms by relieving them of the burden of buying high-cost crude oil on the spot market during a disruption.[7]

During a disruption, a benefitting firm's expectation of oil prices is reduced to the average cost of imports available to large firms. This reduces expectations of profits from stockpiling and should, therefore, reduce stockholding by potential beneficiaries.[8] At the same time, the larger, selling, firms' demand for stocks should not be affected because allocations do not affect the price they expect to

---

7. For instance, in December 1979 Ashland Oil was granted an allocation of 80,000 barrels a day so that it would not have to purchase spot market crude, which was then selling at prices substantially above official crude prices. The allocation was justified by the argument that Ashland had lost a supply of crude from Iran due to President Carter's embargo on purchases of oil from Iran.

8. This can be seen mathematically by letting $P^*$ be the expected price of oil paid by large firms in the event of a disruption and $P'$ be the expected price paid by smaller firms. We assume that $P^* < P'$ because the large firms have contractual relationships with exporting countries that permit them to purchase crude at the official price, while the smaller firms rely on third-party sales or spot market sales at prices which tend to follow spot values. We assume that the demand for inventories is partially determined by the firm's expectations as to future price increases, and that firms will acquire stocks only if

$$E(P') > (1 + r_t) P_t$$

or if

$$E(P') > (1 + r_t)(1 + r_{t-1})(1 + r_{t-2}) \ldots (P_{t-n}).$$

From this rule it is simple to conclude that the smaller firms would tend to acquire smaller stocks if $P^*$ is substituted for $P'$.

receive during a disruption. Thus, total stockholding should be reduced, since allocations reduce the demand for stocks by smaller firms and do not increase the demand for stocks by larger firms.

While price or allocation controls discourage the building of inventories, certain other policies encourage firms to build speculative stocks. The most obvious of these would be a standby disruption tariff that could be imposed on imports of crude oil and product. The tariff would offer the prospect of very substantial, if short-term, profits to speculators in the event of a disruption and ought to induce increased stockholding.

To see how the program would work, assume that a tariff had been used during the Iranian disruption and that price controls had not been in effect. Under these circumstances, a company that had purchased and stored crude before the disruption could have drawn on its stocks during the disruption and sold the oil at the going market price, including the tariff. Since it would not have to pay the tariff on stocks purchased before the disruption, the company would reap a windfall. Thus, if Gulf had acquired oil at $14.00 a barrel in July 1978 and if a tariff of $10.00 had been imposed in January 1979, Gulf could have sold the $14.00 oil for $24.00 in January, even if world oil prices had remained unchanged. In these circumstances the reward for speculating would have been $10.00 a barrel.

Other tax measures, such as refiner taxes or consumer taxes, offer far less incentive, because they require that holders of the speculative stocks incur the burden of taxation in the same manner as the importer or domestic producer of crude oil. The effect on the receipts of speculators, refiners, and crude oil producers may be seen from the following table, which assumes that a tax of $30.00 is imposed at the time of a disruption on top of the predisruption price of crude oil of $30.00.

|  | Price Paid by Consumer | Price Received by Refiner | Price Received by Importer | Price Received by Producer | Price Received by Speculator |
|---|---|---|---|---|---|
|  | *Dollars per Barrel* | | | | |
| Tax on product | $60.00 | $30.00 | $30.00 | $30.00 | $30.00 |
| Tax on refiner | 60.00 | 30.00 | 30.00 | 30.00 | 30.00 |
| Tax on imports | 60.00 | 60.00 | 30.00 | 60.00 | 60.00 |

A final mechanism that would encourage suppliers to increase inventories would be a program analogous to the U.S. system of price supports for farmers. Under such a program stocks would be insulated from the market when prices were low but put back into the market when supplies became short. As described by the President's Council of Economic Advisers (1978) (with respect to wheat), the effect of such a program, if properly managed, would be to provide a wide "corridor" for price fluctuation so that the market could function; to provide protection in the event of a disruption; and not to overhang the market when prices were low.

If agricultural-type price support programs were applied to the oil industry, eligible refiners or speculators could apply for low-interest loans against their stocks. In return, the stocks could not be sold for a period of time unless the price of oil rose above a specific threshold level. If oil prices did rise above that level, the stocks could be sold to liquidate the low interest loans. Furthermore, if prices rose above a second, higher, threshold level, inventories would have to be sold. Thus, if the price support level was established at $20.00 a barrel, the first threshold at $40.00, and the second threshold at $50.00, a firm proposing to borrow against existing stockpiles would be prevented from selling from those stocks until the price rose to $40.00 and would be required to sell from them if prices rose to $50.00 a barrel.

Nordhaus (1980a: 32) endorses this concept, noting, "We encourage stockpiling when prices are low and drawdown when prices are high. Moreover, we have given incentives to farmers and others in the private sector to hold stocks and to dispose of them when profitable—i.e., when prices are high."

The application of the commodity price support program to oil creates two troubling problems. First, farm price supports were designed to provide stability to farmers' income in order to maintain relatively steady rates of production. The oil industry is not likely to be in a situation where it requires a stabilization of income to maintain production. Second, farm price supports work because commodity prices fluctuate rapidly. As noted above, especially in Chapters 2 and 6, one problem with oil markets is that prices do not fluctuate rapidly enough. Thus, while the profitable speculative strategy in agriculture is to sell after a sharp price rise, the profitable strategy in oil is to increase inventories in response to the first surge in oil prices because a second surge is sure to follow.

## Achieving Storage Through Regulation

It is also possible to use regulations to increase stockpiles. Many European countries require oil companies to hold specified volumes of inventories and establish the necessary reporting rules for the distribution of these stocks. For instance, West Germany has developed the Erdoelbevorrantungsverband (EBV), an industrial consortium that manages emergency stockpiles of oil. Membership in the EBV is mandatory for all firms that refine or import oil into West Germany. Companies contribute their oil to the EBV, which arranges for storage. In France, there are specific governmental directives requiring companies to hold inventories, but the inventories are held by the companies themselves (Deese and Nye 1981:197).

The effect of this type of program is to increase the cost of refining and marketing oil. Both West Germany and France recognize this fact, and in France, where price controls are imposed, companies are allowed to pass the cost on to consumers. Since price controls are not imposed in West Germany, there can be no guarantee that the cost of mandatory participation in the EBV can be passed on to consumers. However, the probability of passing on the cost is maximized by making all firms participate in the EBV and by making the degree of participation proportional to the volume of sales in West Germany.[9]

---

9. To demonstrate that the cost should be passed on, we refer to the theory of marginal costs. According to the simplest model of supply and demand, the equilibrium price is determined by the intersection of the industry marginal cost curve and the demand curve. Thus, the amount a firm will supply will be given by

$$q_i = F(P),$$

where $F(P)$ is the marginal cost curve. Now, if a firm is required to contribute a fixed proportion of its sales to a stockpile (or to hold a specific portion of its sales), that proportion may be written as

$$\text{Proportion} = kq_i,$$

where $k$ is the proportion of supplies that must be held in inventories. The cost of holding these inventories is equal to the interest cost, $r_t$, plus any per barrel service charge, $x$, levied by the storage authority. Thus, the firm's total cost would be

$$\text{Cost of storage} = (r_t + x)q_i,$$

and the cost per barrel of supply would be

$$\text{Per barrel cost} = (r_t + x) * k$$

for all firms, so long as $k$ was identical for all firms.

Mandated storage programs such as those imposed by the government of West Germany could be used successfully to build inventories in the United States. Indeed, there should be much less uncertainty as to their effectiveness. However, if improperly applied, such proposals could substantially distort markets during periods of market stability.

Distortion would result if only a portion of the firms in the oil industry were required to hold compulsory stocks because such a bias would give those firms exempted from the storage requirement a large financial advantage. That financial advantage could be translated into lower prices and force the firms required to hold stocks to absorb, rather than recover, the cost of the higher stocks.[10]

Regulations that required only larger firms to hold oil stocks could have a significant effect on the profits of these firms if storage requirements similar to those imposed by West Germany were set.[11] If such requirements were imposed only on the larger firms in the United States and if these firms were unable to raise prices to recover the costs (as might be the case during times of surplus)[12] then the costs of participation for a sample of eleven oil companies in 1980 would have ranged from 10 to 50 percent of 1980 net income.

---

Thus the marginal cost of supplying $q_i$ units would be

$$q_i = F(P) + (r_t + x) * k .$$

This represents an upward shift of the marginal cost curve equal to $(r_t + x) * k$. Since all marginal cost curves would be shifted, the supply curve would be shifted, forcing a portion of the cost of the increased supply on to consumers. The proportion would depend upon the price elasticity of demand. In general, the low elasticity of demand would mean that most of the increase would be borne by consumers.

10. Favorable treatment of one group of firms vis-à-vis another group would have the effect of leaving marginal costs of operation of the favored group unchanged while raising the marginal costs of the unfavored group. Under these circumstances, one of two things would happen: (1) Either the unfavored group would be able to increase prices to recover the cost of storage imposed on them by the central government, which would then allow the favored group to increase market share and profits while the unfavored group lost market share and aggregate profits; or (2) the unfavored group would not be able to increase prices and recover the cost of participation in the storage program, in which case the unfavored firms would pay the cost of storage out of profits. The losses would presumably induce them to restrict sales. On the other hand, the favored firms would not lose profits and could increase their market share.

11. These requirements dictate that oil companies hold the equivalent of 90 days of their prior year's sales in inventories.

12. The likelihood that the costs could be passed on would depend upon the treatment of product and crude imports. If smaller firms could import crude or product without incurring any storage fee, then these imports would tend to set the market price, and the market price probably would not include any recovery of the storage fee.

For instance, the cost to Exxon would have been $617 million, 10.9 percent of net income, while the cost to Marathon Oil would have been $190 million, 50.2 percent of net income. For the eleven companies in aggregate the cost would represent 17 percent of net income. (The full set of estimates is given on Table 9-3.)[13]

Special treatment for one portion of the oil industry is not as farfetched as it may seem. In March 1980 Senator Kassenbaum proposed that companies importing more than 75,000 barrels a day in any calendar year be required to contribute specific volumes of oil to the U.S. strategic petroleum reserve, and Section 156 of the Energy Policy and Conservation Act (EPCA) (1976) authorizes the Secretary of Energy to require some or all importers of oil to set aside oil in an "industrial petroleum reserve." The language of the section specifically directs the secretary to take steps to avoid imposing burdens on smaller firms.

Two criticisms of this approach are offered here. First, as noted above, the exemption of some firms reduces the probability that the costs of storing the oil can be recovered by firms required to hold oil. This economic inefficiency will lead to other distortions. However, the more important failing of such proposals is their isolated focus on imports. Most U.S. proposals specifically single out those firms that import oil, on the theory that firms relying on U.S. production are less vulnerable to a reduction in the supply of oil than are firms relying on imports. This emphasis fails to recognize the essential importance of stocks in reducing the sudden upsurge in prices on commodity markets at the start of a disruption. In summary, then, creation of speculative stocks by regulation represents a workable alternative only if the regulations can be applied equally to all firms, including domestic producers.

### Conclusions on Supplier Inventories

Incentives to build inventories for suppliers appear to be slim in a free market but can be increased by reducing the cost of holding inventories through subsidy mechanisms. A mechanism that reduces

---

13. This calculation assumes that these companies would each have to add stocks equal to 90 days of their 1980 sales. Such an estimate is an extreme case. To the extent that the addition is less than 90 days, the cost is reduced. For example, if the addition were only 30 days for the eleven companies, the cost would only be $1 billion.

Table 9–3. Estimates of Financial Burden to Specific Companies of Carrying 90 Days of Emergency Inventories.

| Company | 1980 United States Sales of Petroleum (Thousand Barrels per Day) | 90 Day Inventory Requirement (Million Barrels) | Cost at 15.2 Percent Interest Rate (Million Dollars) | Net United States Income (Million Dollars)[a] | Cost as a Percentage of Net Income |
|---|---|---|---|---|---|
| Exxon | 1,503 | 135.3 | $ 617 | $ 5,665 | 11 |
| Standard Oil of California | 1,255 | 113.0 | 515 | 2,401 | 21 |
| Shell Oil[b] | 1,012 | 100.8 | 460 | 1,542 | 30 |
| Gulf Oil | 737 | 66.3 | 302 | 1,407 | 21 |
| ARCO | 703 | 63.3 | 289 | 1,651 | 18 |
| Marathon | 464 | 41.8 | 190 | 379 | 50 |
| Phillips | 460 | 41.4 | 189 | 1,069 | 18 |
| Union Oil | 421 | 37.9 | 173 | 647 | 27 |
| Sohio | 413 | 37.2 | 169 | 1,811 | 9 |
| Conoco | 338 | 30.4 | 139 | 1,026 | 14 |
| Cities Service | 270 | 24.3 | 111 | 478 | 23 |
| Eleven company total | 7,666 | 691.7 | 3,154 | 18,076 | 17 |

a. World wide income as reported in annual reports net of all taxes and expenses.
b. Shell Oil Co., USA.

the interest cost of holding inventories offers the best incentive because it is paid while oil is held in stocks, not promised for some date in the future.

As an alternative, increases in prices paid for oil during a disruption would also offer an incentive to build stocks. The most positive type of incentive would be a tariff on imported oil. Price controls, on the other hand, destroy all incentives.

## INCENTIVES TO BUILD CONSUMER INVENTORIES

While most discussion of stock management has focused on the inventory management practices of the oil industry during past disruptions, the inventory practices of consumers and their possible participation in the development of standby stocks represent a second major source of emergency supply. Standby stocks held by consumers offer the same benefit to the economy in moderating increases in oil prices during disruptions as speculative stocks held by suppliers. However, the word speculative is a misnomer in the case of consumer inventories—a more appropriate term would be "insurance" or "precautionary" stocks.

The size of existing consumer stocks and the potential to increase them is unknown because, with the exception of the utility sector, no data on consumer stocks are collected. However, the incentives to increase storage capacity and stocks can be identified. Indeed, they are identical to those facing suppliers of oil.

The basic rule for storage applies to consumers just as it applies to suppliers. Consumers should stockpile product if the expected price in a future period is greater than the current price multiplied by one plus the market rate of interest.

Unfortunately, the application of this rule during the 1970s would have led to the same behavior predicted of suppliers: liquidation during times of plenty and acquisition during times of shortage. Furthermore, price control regulations and other regulatory programs, particularly end-use allocations, reduced the incentive to stockpile.

Adoption of aggressive pricing strategies or announcement of intentions to impose disruption taxes would, on the other hand, increase the incentive to build stockpiles. Deese (Deese and Nye 1980: 196) reports that consumer inventories of heating oil are very large in West Germany, where aggressive pricing would be used.

The discussion of alternative policies to encourage stockbuilding by suppliers presented in the previous section applies generally to consumers. Specifically, the prospect of imposition of price controls will discourage stockbuilding, while the prospect of imposition of a large disruption tariff will encourage stockbuilding, as will tax or subsidy incentives. Two of the programs will, however, have slightly different effects: These are standby taxes and product allocation programs.

From the point of view of the consumer, standby taxes and standby tariffs have identical effects. Both increase the price the consumer expects to pay during a disruption, and both should encourage consumers to hold increased stocks. On the other hand, standby tariffs would encourage greater holdings of inventories by suppliers than standby taxes.

Product allocation programs also may have a different effect on consumers than crude oil allocation programs have on suppliers, because product allocation programs historically have been structured differently than crude oil allocation programs. Crude oil allocation programs assure smaller refiners a relatively stable supply of crude oil at stable prices, thus discouraging stockbuilding. Product allocation programs, on the other hand, do not offer the consumer the same type of guarantee. Instead, these programs historically have established a system of priorities that a distributor was required to follow. From 1973 to 1981 farmers, fire departments, and other emergency services were entitled to receive 100 percent of their requirements of gasoline; industrial users such as trucking companies were limited to 100 percent of their base period uses, and allocations to other consumers, such as the average motorist, were made under the "crumb theory" — that is, they received whatever was left over.

The economic effect of this type of allocation system on consumers would depend upon the consumer's status as defined by the regulations. While those consumers who received priority treatment would not be expected to build precautionary stocks, those who received low priorities might be expected to purchase tanks and other oil related storage equipment in order to improve their preparations for dealing with a disruption.[14] The limited amount of anec-

---

14. In terms of the storage rule, one can view the effect of allocations as setting up a two-tiered pricing system during a disruption. Priority customers are able to buy at low prices and thus have no incentive to build stocks, while low priority customers must pay an excessive price (perhaps even an infinite price) for incremental supplies and thus have a strong incentive to stockpile.

dotal evidence from the embargo period suggests that this is precisely what happened.

## BUILDING PUBLICLY OWNED STOCKS

The individual governments of the most developed consuming countries have concluded that the size of private stocks is insufficient and have agreed to develop additional stocks beyond those that would be held privately. According to Deese (Deese and Nye 1981), the International Energy Agency (IEA) agreement requires each member country to hold stocks equal to ninety days of net imports.[15] The United States is required to hold 700 million barrels in storage. The volume of oil that must be held in federally owned stockpiles is less because the IEA allows commercial stocks to be credited against the 700 million barrel target.

The actual targets established by the governments of most countries exceed the IEA inventory targets. As Deese (Deese and Nye 1980:136) notes, the governments of West Germany, France, Italy, and the Netherlands have each inaugurated specific programs to assure that emergency inventories, as distinguished from commercial inventories, are in excess of ninety days of consumption. In fact, as a recent report by the National Petroleum Council noted, only Canada and New Zealand, of all the IEA countries, have no program at all, and only Canada, New Zealand, Switzerland, and the United States do not set minimum industry inventories. (See Table 9-4.)

The justification for building public stockpiles rests on an assumption that the stocks held by the public augment private stocks and can be used to moderate price increases in a disruption. This justification rests on two premises: first, that public stockpiles augment private stocks and, second, that public stocks will be used. Both assumptions may be incorrect.

This problem was recognized by the National Commission on Supplies and Shortages in a report issued in December 1976. The commission examined the reasons why public stockpiles might be required and concluded that private stockpiling would not be ade-

---

15. The members of the International Energy Agency are the United States, Canada, Japan, Australia, New Zealand, West Germany, Italy, Great Britain, Denmark, Sweden, Austria, Belgium, Greece, Ireland, Luxembourg, the Netherlands, Norway, Spain, Switzerland, and Turkey.

Table 9-4. Oil Stockpiling Programs of Eighteen IEA Countries.

| Country | Import Dependence[a] (percent) | Government Reserve | Public Corporation | Industry Minimum Storage | No Program |
|---|---|---|---|---|---|
| Australia | 27 | | | x[b] | |
| Austria | 86 | | | x | |
| Belgium | 100 | | | x | |
| Canada | 8 | | | | x |
| Denmark | 97 | | | x | |
| West Germany | 96 | x | x | x | |
| Greece | 100 | | | x | |
| Ireland | 100 | | | x[c] | |
| Italy | 98 | | | x | |
| Japan | 99 | x | | x | |
| Luxembourg | 100 | | | | x |
| Netherlands | 93 | | x[d] | x | |
| New Zealand | 92 | | | x | |
| Spain | 97 | | | | |
| Sweden | 100 | x | | x | |
| Switzerland | 100 | | x | x | |
| United Kingdom | 18 | x | | | |
| United States | 43 | x | | | |

a. OECD quarterly oil statistics, 1980: 1.
b. As of mid-1980, stocks were at sixty-seven days' consumption; the government has announced plans to raise this to seventy-five days' worth.
c. Italy requires major consumers to maintain a minimum storage fill equivalent to 20 percent of their storage capacity.
d. The Netherlands program is in transition from an oil-company run to a public-corporation run strategic storage program.

Source: National Petroleum Council, *Emergency Preparedness of Interruption of Petroleum Imports into the United States*

quate in cases where either the existence of a stockpile would not deter formation of a cartel or subsequent manipulation of prices or in cases where there was a possibility that price controls might be imposed in the event of a shortage. In these cases (which are clearly cases that apply to oil), the development of public stockpiles was warranted. The commission (National Commission on Supplies and Shortages, 1975:128) noted, though,

> Increasing the size of the public stockpile (perhaps from zero) will probably reduce somewhat the level of private stocks held in anticipation of future supply disruptions. The size of the reduction depends in part on the purpose of the government stockpile and the management of its acquisition and sale of stocks. If the stockpile agency acts to maximize the expected profits from stockpiling (which is similar to buffering prices), private firms may view government stocks as perfect substitutes for their own stocks; in that case an increase in government stocks of one unit will lead private firms to decrease their stocks by the same amount.
>
> On the other hand, if the government stocks are to be used only in very severe shortages, or in cases thought by the private sector to be quite unlikely, the reductions in private stocks will be much less. Economic theory suggests that a two-unit increase in government stocks may be accompanied by as much as a one unit decrease in private stocks.

In an assessment of the economic effect of a strategic petroleum reserve on private inventories, Wright and Williams (1981) concluded that the creation of a reserve would reduce the level of private storage. Wright and Williams reached this conclusion through 1,000 simulations of a dynamic programming model in which the probability of a disruption in any period was specified.[16]

The results of these simulations indicate that private speculative (as opposed to working) inventories were substantially reduced under a regime in which price controls were imposed as opposed to one in which no price controls were imposed and that private speculative inventories would be cut further if a government reserve were created.[17]

---

16. Wright and Williams assume that there is a 15 percent chance of a 15 percent cut in imports in any year, a 10 percent chance of a 30 percent cut in any year, and a 5 percent chance of a 45 percent cut in any year.

17. The Wright/Williams results indicate that private firms would hold 780 million barrels of speculative inventories (i.e., inventories not included in working levels) if no price controls were imposed but only 240 million barrels if controls were imposed. In both cases they assume that there is no public storage. When public storage is added to the price control simulation, they find that private inventories are further reduced from 240 million barrels to 190 million barrels, while the optimal public storage is 710 million barrels.

Unfortunately, however, the results are only suggestive because Wright and Williams are forced to make several simplifying assumptions to make their model function. The most critical of these assumptions are that prices adjust instantaneously, that prices decline after a disruption is over, and that public inventories are acquired and liquidated in response to the short-term movement of prices. All three assumptions are absolutely necessary to apply the approach used in the study, which is based on the theory of intertemporal arbitrage. While the theory is sound, the conditions of the oil market violate the assumptions, as indicated in Chapters 2 and 3.

Since the Wright/Williams study represents the only effort to analyze the effect of a public stockpile on private stockholdings of oil, the assessment of the effect of the SPRO on private stocks must be summarized with a very weak conclusion that public stocks may affect the level of private stocks but that the degree of the impact is not known. There might be no impact in the extreme case where consumers and suppliers believed the reserve would never, or could never, be used (as might be the case if the bureaucracy managing the oil was viewed as hopelessly moribund.) On the other hand, if market participants believed that the reserve would be managed aggressively, the conclusions of the National Commission on Supplies and Shortages would apply.

# 10 POLICIES TO REDUCE CONSUMPTION DURING A DISRUPTION

> The conference furthermore expressed its deep concern for the lack of necessary measures that should be taken by the industrialized developed countries with a view to controlling the market situation.
>
> Exerpt from OPEC communique issued in March 1979 during the Iranian crisis. *Petroleum Intelligence Weekly* (April 12, 1979:11)

Consumers and oil consuming countries have been criticized frequently for profiligate consumption of energy and oil. For instance, Jimmy Carter and James Schlesinger both blamed U.S. consumers for failing to conserve, and the editors of *The New York Times* rarely allow a week to pass without hitting American consumers for their antisocial patterns of energy consumption. These views have been echoed by the ministers of the OPEC nations who, as the quote above indicates, lay the blame for higher oil prices with consumers.

During periods of stable markets, such accusations are nonsense, because the amount of energy consumed is determined by economic conditions and prices. However, the argument takes on an element of truth during periods when markets are disrupted. During such times, collective action by consumers can moderate the rate of increase in the price of oil.

The effectiveness of these measures depends upon several factors: the speed with which they are introduced, the size of the reduction in consumption, and the length of time the measure can be sustained. A measure that can be imposed quickly is more valuable than one that requires time to be started. Thus, one measure that induces only a small immediate reduction in the consumption of oil may be preferred to one requiring several months to implement.

It is also important that a measure be sustainable over a period of time. Measures that require continued public awareness tend to be unsustainable because public interest may wane before the disruption has passed. As attention drops, demand picks up, causing greater price pressure.

The effectiveness of various emergency conservation measures will be examined by presenting a review of the various measures that have been proposed. The effectiveness of these measures in reducing pressures on prices will then be evaluated. The latter evaluation will be made by simulation using the model of oil market behavior.

## EMERGENCY MEASURES FOR REDUCING CONSUMPTION

The list of possible emergency measures that have been proposed to reduce oil consumption is long, but the number of such measures that have a chance of inducing meaningful reductions in consumption during an emergency is quite small. Many of the proposals are so complicated that they simply will not work, while others are so sensible that they already have been adopted as cost-saving measures. Other measures, like gasoline rationing, require such long start-up times that they would be ineffective because they could not be put in place in time to prevent a price spiral.

The proposals reviewed here are drawn from materials issued by the Department of Energy and its predecessor agencies during the 1973 and 1979 crises, as well as from the recent report of the National petroleum Council (NPC) (1980), *Emergency Preparedness for Interruption of Petroleum Imports into the United States*. Where possible, these proposals are also compared with programs that have been implemented in Europe. The policies examined are:

- Building temperature controls;
- Natural gas or coal substitution;

- Electricity wheeling;
- Other electricity savings;
- Personal fuel conservation;
- Conservation in the transportation sector; and
- Gasoline rationing.

### Building Temperature Controls

Building temperature controls or, more precisely, the reduced use of oil for heating and cooling, have been a central element in each emergency conservation plan developed in the past and are essential elements of current emergency conservation plans in both Germany and France. (Temperature control plans presumably would be part of a U.S. emergency plan if such a plan existed.)

The Nixon administration relied heavily upon reductions in temperatures for buildings heated by distillate oil when it hastily formulated its emergency conservation plan in 1973. This plan (White House Press Office Nov. 25, 1973: 5) stated, "Reduction of residential heating by 6° and commercial heating by 10° saves 490,000 barrels per day."[1] These reductions were to be achieved by mandatory end user allocations of heating oil. The plan (White House Press Office Nov. 25, 1973: 5) stated that regulations would be proposed to "reduce amounts available for residential space heating by 15 percent, commercial and other space heating by 25 percent and industrial uses by 10 percent." The program was imposed during December 1973 and January and February 1974.

The Carter administration (U.S. Department of Energy 1979) proposed a similar program at the time of the Iranian disruption. Using authorities granted under the Energy Policy and Conservation Act, the administration required that commercial building (offices, schools, and other public buildings) temperatures be no warmer than 65° during the winter and no cooler then 78° during the summer. Whereas the Nixon administration regulations had applied only to consumers of heating oil, the Carter regulations applied to buildings heated by oil, gas, or electricity, on the theory that the natural gas conserved by such actions could then be used to substitute (or "back

---

1. The Fact Sheet uses the word "residual" rather than "residential." However, this was apparently a typographical error.

out") oil in another activity. These regulations were imposed in April 1979 and then remained in effect for the purposes of "encouraging conservation" until January 1981 when they were repealed by President Reagan. The Carter administration estimated that adherence to the temperature regulations would reduce consumption by 150,000 to 300,000 barrels a day.

The emergency energy programs adopted by West Germany and France (Deese and Miller 1981) are similar to those used by the United States in 1973 and 1979. In West Germany, regulations stipulate that in a major emergency fuel oil dealers would make a quarterly allocation of fuel oil to consumers that would represent a fixed percentage of historical consumption. In France, consumption of heating fuel in homes and offices has been limited to 90 percent of 1979 levels since 1979. This percentage would presumably be reduced in case of an emergency.

The popularity of building temperature regulations as emergency measures in the United States and Europe can be traced to their ease of imposition, their acceptance by the public, and their minimal effect on the economy. However, the potential contribution of these measures to reduce oil consumption may be problematical.

The principal advantage of temperature controls is that they can be imposed quickly. At a time of a recognized national emergency it is fairly simple for the President or head of state to call on the citizenry of a nation to turn thermostats down (or, during the summer, up). Furthermore, the request is enforceable so long as the crisis remains in the public view through the use of peer pressure on those who fail to comply. During a crisis it becomes unseemly to overheat or overcool a building frequented by the public. Another advantage of building temperature controls is that they reduce the consumption of energy without greatly affecting economic activity. Thus, supplies of oil are preserved for productive activities.[2]

The principal drawback with building temperature proposals is that they offer only minimal savings. As we show on Table 10-1, the

---

2. Deese (Deese and Miller 1981: 196) noted that the German plan is predicated on this principal. "Consistent with West Germany's free market approach to energy, Type I crises are expected to require only limited government intervention. Government agencies would meet with consumers and provide policing functions as necessary, but they would avoid formal bureaucratic action. The government would intervene much more actively in a Type II crisis, however, in order to maintain the productive sectors and continue the delivery of such essential services as fire, police and hospital operations."

**Table 10-1.** Estimates of Potential Conservation by Building Temperature Controls.

|  | Savings | | | |
| --- | --- | --- | --- | --- |
| Estimate | Oil | Gas | Electricity | Total |
|  | Thousands of Barrels a Day | | | |
| White House (1973a) | 490 | | | 490 |
| U.S. Department of Energy (1979)[b] | | | | 150-300 |
| National Petroleum Council (1981) | 50 | 105 | 10 | 165 |

a. All savings achieved by reducing use of heating oil in homes, offices, and factories where heating oil was consumed.

b. Restrictions apply to buildings heated by all types of fuel. It is not clear, however, whether the savings estimate represented savings of oil alone, or of oil, gas, and electricity.

NPC data now project that savings from these measures in the United States would amount to only 50,000 barrels a day (bd) of oil, the equivalent of 105,000 bd of gas, and perhaps 10,000 bd of electricity. These small savings contrast with the much higher estimates made by the Carter administration (150,000 bd to 300,000 bd) and the Nixon administration (490,000 bd).

The differences between the Nixon, Carter, and NPC estimates can be traced to two factors. First, both the Carter and Nixon estimates were made at a time of crisis when there was no time to make detailed studies, while the NPC estimates were made as part of a deliberate, careful study when there was no emergency. Second, as the NPC notes, many of the potential savings that were once available to the economy are no longer available because consumers have made permanent adjustments in their patterns of consumption. This is by far the more important explanation for the discrepancy. The NPC (1981:88) notes,

> "Previous sources indicated that 8 to 10 percent of space conditioning energy could be saved by reducing thermostat settings to 68° in the winter and raising them to 78° in the summer. However, at least half of this potential has already been achieved and the realization of another 20 percent is considered doubtful.

The implication of the NPC assessment is, then, that while building temperature controls offer a source of quick conservation, the potential volume of conservation is not all that great. Furthermore, as we note on Table 10-1, most of the conservation occurs due to the reduced use of gas, not oil. Thus, for building temperature controls to result in a reduction in oil use, it would also be necessary to identify consumers of oil who can switch to gas.[3]

### Fuel Substitution Measures

A second popular emergency measure for dealing with a disruption is fuel switching. Many plants and large fuel burning institutions have the capacity to burn more than one type of fuel and can change from one fuel to another on fairly short notice. For instance, many plants constructed during the late 1950s, 1960s, and early 1970s were designed to operate on either natural gas or oil, with the preferred fuel being natural gas because of its lower price and cleaner burning characteristics. Some electric utility plants were also originally built to burn coal and then later modified to burn oil.

U.S. emergency conservation policy during the disruptions in both 1973 and 1979 attempted to make the maximum use of this fuel switching capability. In 1973, the Nixon administration indicated that it hoped to reduce oil imports by displacing 250,000 bd of oil with coal and another 150,000 bd with gas. The increased use of coal was to be achieved by conversion of oil burning electrical generation plants to coal, while the increased usage of gas was to be realized by "reducing residential consumption of natural gas so that electric utilities can use it instead of residual oil for power generation" (White House Press Office Nov. 25, 1973: 5).

The available statistics suggest that, while logically formulated, this plan was unsuccessful. Indeed, the newspaper reports during the embargo period suggest that utilities were unable to purchase either the coal or gas.

The *Iranian Response Plan* (U.S. Department of Energy 1979) developed by the Carter administration also called for substitution of gas and coal for oil. The plan envisioned increasing the use of gas by

---

3. The need to induce savings of non-oil fuels and also to argue for the non-oil fuels to be substituted for oil in other uses is often overlooked in studies of energy conservation.

the equivalent of 250,000 to 400,000 bd. This was to be achieved by pushing gas on industrial customers who had dual fired plants. This program was probably more successful because it occurred at a time when natural gas was in surplus (in 1973 natural gas was in short supply due to the effects of nineteen years of interstate price regulation) and because recently passed legislation pertaining to the use of fuels in large boilers (the Fuel Use Act) had started a process of conversion away from gas. This process was easily reversible.

The National Petroleum Council's 1981 analysis estimates that conversion to natural gas could reduce oil consumption by as much as 1 million barrels a day in the event of a disruption, while conversion from oil to coal might reduce industrial use of oil by 50,000 bd. This is a much larger savings than was projected by DOE in 1979. The increased use of gas would be divided between utilities (700,000 bd) and industrial users (260,000 bd).[4] The increased use of natural gas by utilities would not, however, lead to displacements of a full 700,000 bd of oil for two reasons. First, some of the utility units that could be switched from oil to gas probably would be idle at the time of a disruption because utilities would be using more efficient nuclear or coal fired plants. Second, some of the dual fired plants have already been granted exemptions from the fuel use act and are already burning gas. Thus, the NPC estimates that total actual reduction of oil use in the utility sector due to the switch to gas would only be 240,000 bd.

The NPC is quite pessimistic about the potential shift from oil to coal in the utility sector. The Council notes that pollution variances would be required in some cases and that the plants that are candidates for conversion generally are not located in the areas where utilities burn oil. Thus, power would have to be "wheeled" from coal consuming utilities in regions where oil and gas are burned. These are located on the East and West Coasts.

The potential for wheeling is limited. The estimates from the three different assessments are compared on Table 10-2. Note that the more recent estimates are larger than earlier estimates.

---

4. The NPC notes that the American Gas Association has estimated that industrial fuel switching could reach 700,000 barrels a day. However, the NPC (1981:93) argues that less than 50 percent of the oil used in the industrial sector could be shifted to gas or would result in savings of oil use if shifted to gas use.

**Table 10-2.** Estimates of Oil Savings Available by Fuel Switching.

| Study | Savings Realizable By Increased Gas Use | | | Savings Realizable By Increased Coal Use | | |
|---|---|---|---|---|---|---|
| | Industry | Utilities | Total | Industry | Utilities | Total |
| | | | Thousand Barrels a Day | | | |
| White House Press Office (1973a) | | 100 | 100 | | 250 | 250 |
| U.S. Department of Energy (1979) | | | 250-400 | | | 0 |
| National Petroleum Council (1981)[a] | 260 | 240-700 | 500-960 | 50 | 30 | 80 |

a. The NPC finds that utilities have the capacity to convert as much as 700 mbd of oil consumption to gas. However, constraints on the demand for electricity generated by oil plus conversions which have already taken place limit this saving to 240 mbd.

## Electricity Wheeling

The wheeling of electricity is a process by which fuel consumption by a utility in one region is replaced by fuel consumption by a second utility in another region. The electricity produced by the second utility is then transferred to the first utility through the elaborate electricity intertie system. Through this system it is possible for a utility in New York to close down one or more of its oil fired boilers and purchase electricity generated by burning coal from a utility in Ohio.

During the Iranian crisis, the Carter administration encouraged utilities to take advantage of the possibility of electricity transfers through power pools in order to displace oil use. In the *Iranian Response Plan*, DOE (1979) estimated that such exchanges could reduce oil use by between 100,000 and 200,000 barrels a day. DOE also suggested that it would resort to powers granted under the Federal Power Act to order transfers to achieve such savings if the exchanges were not arranged voluntarily.

The National Petroleum Council's 1981 report offers a much more pessimistic assessment, suggesting that the maximum savings from wheeling might amount to only 30,000 bd. The large discrepency between the NPC and the 1979 *Iranian Response Plan* estimates can be traced to two factors. First, the chief vehicle for wheeling is the power pools, which are "formal organizations established by two or more utilities to improve service reliability through joint planning and operation (National Petroleum Council 1981:93). NPC notes that the purpose of power pooling is efficiency, not provision of surplus transmission capacity to be used in an emergency, and states that a very limited amount of surplus capacity is available.

A second reason for the NPC's (1981:93) pessimism is the fact that the power pools are already being used to reduce oil use. "The DOE data do demonstrate that power pooling is currently used by utilities to reduce oil use by approximately 300 to 350 mbd. Oil displacement in 1980 appears to have increased over 1979 levels by 50 mbd." In other words, because power wheeling is such a good idea, firms are already practicing it. (See Table 10-3.)

### Other Measures to Reduce Oil Use in Electricity Generation

Measures other than wheeling have also been proposed to reduce the use of oil in the electric utility sector. The emergency conser-

Table 10-3. Estimates of Oil Savings Through Power Wheeling (*thousands of barrels a day*).

| Study | Oil Savings | Comment |
|---|---|---|
| White House Press Office (1973a) | No estimate | |
| U.S. Department of Energy (1979) | 100-200 | 100 mbd listed as most likely figure |
| National Petroleum Council (1981) | 50 | Wheeling already taking place |

vation plan announced for the 1973 embargo called for a 3 percent cut in electricity consumption. It was estimated that such a cut would reduce the use of oil in the utility sector by 300,000 barrels a day. This saving was presumably to be achieved by direct reductions in oil consumption by those utilities using oil to generate electricity, and possibly by further reductions of oil use by the same utilities to the extent that reduced consumption of electricity in other regions reduced the load on power interchanges and permitted increased wheeling.[5]

The 1973 emergency plan also projected reduced utility use of oil due to substitution of natural gas for oil in dual fired plants. The incremental supply of gas was created by conservation in the residential sector.

Neither the 1979 *Iranian Response Plan* nor the 1981 NPC study takes into account this type of potential second-round saving envisioned by the 1973 study. Thus, in the *Iranian Response Plan* there is no suggestion that the surplus gas or coal created in one sector by conservation (such as temperature controls) might find its way to a second sector. Nor is there any attempt in the NPC study to measure the increased wheeling potential that might arise if levels of electricity were reduced.

The NPC study does, however, identify four other actions in the utility sector that might reduce oil use: increased electricity imports from Canada (this would save 30,000 barrels a day); a 5 percent cut in voltage (this would save 20,000 bd); reduction of electricity demand (estimated savings would be 50,000 bd); and increased production from nuclear power plants (30,000 bd).

Of the four measures suggested by the NPC, the demand reduction measure probably has the greatest chance of success. The NPC (1981:77) notes that the city of Los Angeles had great success in cutting electricity consumption during the 1973 embargo when the local municipal power company was forced to severely curtail output. To achieve the reduction in consumption, Los Angeles passed a municipal ordinance mandating cuts in electricity use of 10 percent in industry and residences and 20 percent in commercial buildings.

---

5. This feature of the 1973 plan highlights one oversight in the 1981 NPC study. Specifically, the small incremental wheeling capacity identified by the NPC assumes that levels of consumption of electricity remain unchanged. However, if electricity consumption is reduced as part of an emergency conservation measure, then the decline in consumption will permit utilities using coal or nuclear plants to wheel more power to oil based utilities.

Total consumption was to be cut by 12 percent. The response of the citizens was far greater than was expected. Total consumption fell 22 percent below predicted levels.

The experience in Los Angeles may be transferable to the rest of the nation, although the NPC report cautions that exactly identical results should not be expected because the mix of consumers in Los Angeles includes more commercial and residential establishments, which can cut more easily, and fewer industrial users. Furthermore, the NPC report notes that the substantial conservation in electricity use since 1973 probably reduces the potential for future conservation.

However, even if the potential is much smaller, a program of nationwide emergency electricity conservation combined with maximum usage of the transmission facilities made surplus by the conservation probably would induce a reduction in electricity far greater than the 50,000 barrels a day estimated by the NPC. In fact, the savings could reach 300,000 to 400,000 barrels a day during a severe disruption if a nationwide effort were made to reduce electricity consumption by 7 to 8 percent.[6]

In summary, then, the utility sector may offer the potential for greater savings than have been estimated in prior reports. The Los Angeles experiment certainly seems to justify this conclusion.

### Reduction of Personal Vehicle Use

In every crisis energy planners have looked to the personal vehicle to provide the greatest reductions in oil use. In both the 1973 and 1979 crises, major attempts were made to reduce the personal use of automobiles, and during each crisis debate over gasoline rationing generated considerable controversy.

This experience in the United States was by no means unique. France, West Germany, and the Netherlands all adopted draconian

---

6. This calculation is based upon the following considerations. Oil accounts for roughly 10 percent of all energy consumed by utilities. Thus, a 1 percent reduction in electricity consumption could, if entirely met by reduced oil use, cause a 10 percent reduction in the quantity of oil consumed. Since the utility industry consumes roughly 1.1 million barrels of oil a day, a 10 percent saving would amount to approximately 100,000 barrels a day. Extending this analysis to a 6 percent cut in electricity consumption, one could conclude that the savings in oil use would amount to 600,000 barrels a day. Assuming that only half of the saving could be realized by wheeling reduces this estimate to 300,000 barrels a day.

measures to reduce gasoline use during the 1973 embargo. In some countries, Sunday driving was prohibited, while in others sales of gasoline were curtailed.

In the United States many measures have been proposed, but few have been implemented because it has been impossible to achieve any degree of consensus among the various constituencies. Instead, transportation fuel use has been curtailed by limiting available supplies, and allocation has been achieved by queuing. Among the proposed measures that are identified in the NPC study are:

- Reductions in speed limits;
- Reductions in personal vehicle use;
- Increased use of car pools;
- Driver education;
- Intensive vehicle inspection to improve efficiency;
- Supplementing public buses with the use of school buses;
- Limitation of sales by license plate numbers (odd-even plans);
- Limitations in the number of days a vehicle may be used (sticker plans);
- Shorter work weeks;
- Limitations on recreational uses of vehicles; and
- Bans on weekend sales of gasoline.

(These items are listed in approximate order of their projected effectiveness in reducing personal fuel consumption.)

While each of these measures would achieve some savings in personal oil use, the greatest savings are thought to be achievable through gasoline rationing. In most cases, however, the savings are entirely hypothetical because it has been impossible to develop a political consensus to support any of the proposed measures. Three examples—recreational uses of fuels, weekend closing of gasoline stations, and gasoline rationing—illustrate the problems.

A ban on the recreational use of gasoline and diesel fuel was proposed by both the Nixon administration in 1973 and the Carter administration in 1978 (when it submitted contingency emergency energy plans). Both plans met strong opposition in Congress from the owners of marinas and other establishments that provided services to owners of pleasure boats. As a result of the opposition, the plans were withdrawn, and no limitations on the recreational use of fuels were imposed.

Proposed weekend closings of filling stations met a similar fate. The emergency plan proposed by President Nixon requested authority to order filling stations to close on the weekends on the theory that this would limit the use of vehicles for pleasure. A similar request was made by President Carter in 1978 when standby conservation programs were proposed. Once again, the efforts were thwarted by representatives of businesses that benefit from weekend driving.

A different form of political coalition defeated a proposed gasoline rationing plan in the spring of 1979. This plan, submitted to Congress by the Carter administration in order to fulfill specific requirements dictated by the Energy Policy and Conservation Act, initially stipulated that every auto would receive an identical number of marketable ration tickets, regardless of the state of registration. However, the administration was forced by the Senate to amend the plan to vary the allocation depending upon the vehicle's state of registration so that autos registered in rural states would receive more ration tickets than those registered in urban states. This amendment then caused members of the House of Representatives (a majority of whom come from urban states) to reject the plan.

The discussion of these three attempts to impose emergency conservation programs that cut back on personal use of vehicles indicates the problems that the federal government has experienced when it has attempted to single out the personal automobile as a source of emergency conservation. This analysis of the history leads to the conclusion that curtailment of personal transportation use of gasoline represents a very difficult means of cutting back on the consumption of petroleum.

Despite these problems, every emergency plan has made reductions in personal use the largest single emergency conservation measure because consumption of gasoline and diesel fuel represents such a large portion of total U.S. petroleum use. The 1973 Nixon plan called for a reduction in gasoline use of at least 1.1 million barrels a day (24 percent of estimated precrisis consumption). In 1979 the Carter administration forcast reductions of from 250,000 to 400,000 barrels a day of gasoline through voluntary measures while calling on Congress to give it the authority to order the closing of gasoline stations on weekends to save a further 110,000 to 270,000 barrels a

Table 10-4. Estimates of Reductions in Personal Fuel Use by Three Different Studies.

| Means of Savings | White House Press Office (1973a) | Study U.S. Department of Energy (1979) | National Petroleum Council (1981) |
|---|---|---|---|
| | Thousand Barrels a Day | | |
| Closing stations on weekends | 50 | 120-235 | 240 |
| Speed limit enforcement (55 mph)[d] | 200 | – | 90 |
| Car pooling | | – | 380 |
| Reduce personal vehicle use | | | 50 |
| Reduced recreational use | | | 30 |
| Other personal saving[a,b,c] | 850 | 220-400 | 215 |
| Odd/even sales | | | 145 |
| Sticker plans | | | 50 |
| Four day work week | | | 35 |
| | 1,100 | 340-635 | 1,235 |

a. The Nixon plan calculated a reduced supply of gasoline equal to 1.4 million barrels a day and arbitrarily assumed that 1.1 million barrels a day would result from reduced personal use. The 850 estimate is derived by subtracting identified savings from the total of 1.1.

b. The Carter administration attributed the savings of 220 to 440 mbd to a variety of "voluntary" personal measures, including adherence to the 55 mph speed limit, curtailment of pleasure driving, pleasure boating, and so forth. The savings were not attributed to individual activities.

c. Includes driver education, vehicle inspection, and school bus utilization.

d. The savings from imposition of speed limits are not comparable because the speed limit in most states in 1973 was substantially in excess of 55 mph.

day. Thus, the total potential savings identified by the Carter administration could have been as high as 650,000 barrels a day. The NPC study indicates that gasoline consumption might be cut by as much as 1.2 million barrels a day through the imposition of a variety of measures. The details of the estimated savings are shown on Table 10-4.

## Emergency Conservation in the Transportation Sector

Transportation uses account for approximately 50 percent of all petroleum consumed in the United States and for an even higher percentage of oil consumed in other developed countries. Over half of the petroleum consumed within the transportation sector is used for personal purposes. The remaining portion, which may vary from 40 to 50 percent of total use, is utilized by commercial enterprises for the purpose of moving goods, people, or energy. These areas offer a potential for reductions in consumption in emergencies, although the institutional structure of the transportation industry occasionally makes it difficult to realize the reductions.

The emphasis in most emergency conservation plans has been on a redirection of movements of goods and people away from forms of transportation thought to use oil intensively (trucks and airplanes) and to other less intensive users of oil (trains, barges, and buses). Thus, in the 1973 plan, fuel allocations for general aviation purposes were cut by 40 percent, while fuel for allocations for "high priority aviation such as air taxi services" were only cut by 20 percent. Allocations of jet fuel to airlines were also reduced by between 5 and 15 percent, and airlines were encouraged to offset reductions in the supply of fuel by increasing load factors and by flying slower.

The 1973 plan also targeted trucks and buses for reductions in fuel use through the imposition of a national 55-mile-per-hour speed limit. At the same time, President Nixon's energy advisers called on the Interstate Commerce Commission to lift back haul restrictions and to allow truckers to follow the most direct routing. These were regulatory actions that would increase the utilization rate of capital and thus reduce energy requirements.

Since 1973 some of the emergency regulatory changes proposed for dealing with the 1973 crisis have been made as part of the effort to eliminate regulation throughout the economy. For instance, the Motor Carrier Act of 1980 increases the flexibility of transportation.

On the other hand, deregulation may also reduce the government's ability to achieve reduction in fuel use in some transportation sectors. Deregulation of the airlines, for instance, while economically efficient, also would eliminate the Civil Aeronautics Board (CAB) and thus would eliminate the agency that enforces cutbacks in airline schedules designed to increase load factors during a disruption.[7]

Of course, reductions in fuel use by truckers and airlines can be achieved through "starvation"—that is by limiting the amount of fuel available through allocations. Such mechanisms, however, are cumbersome, inefficient, and, in the case of the trucking industry, ineffective, as was observed in both 1974 and 1979.

In summary, emergency fuel conservation measures in the commercial transportation sector are difficult to identify and even harder to achieve. Deregulation will make it harder, not easier, to achieve mandated reductions in fuel use. Published reports (National Petroleum Council 1981) estimate fuel savings in the commercial transportation sector on the order of 100,000 barrels a day. These savings would result from increased airline seat load factors and elimination of general aviation flights. In addition, the NPC estimates that savings also could be achieved by relaxing CAB rules, improving flight operations, and voluntarily reducing military flights. However, no estimates of the savings from these measures are given here.

### Gasoline Rationing

The final emergency conservation measure proposed for meeting a disruption is gasoline rationing. Rationing programs were proposed during both the 1973 and 1979 disruptions, and in 1973 the federal government even went so far as to order coupons printed. However, for several reasons no rationing mechanism has been implemented.

---

7. During the late 1960's and early 1970's the CAB permitted airlines to enter into capacity agreements. Under these admittedly anticompetitive arrangements, airlines were permitted to allocate the share of the market among themselves. While these arrangements were in effect, the practice of scheduling competitive departures declined. (Competitive departures occur when two airlines schedule flights to the same destination leaving at approximately the same time.) Load factors increased. The antitrust division of the Justice Department naturally objected to these programs, as did most economists, and airline deregulation legislation passed in 1978 banned their future use.

The airline deregulation legislation also contemplates the elimination of the Civil Aeronautics Board by 1983. Once the CAB is eliminated, there will be no formal agency to impose increases in load factors. However, the Federal Aviation agency could enforce conservation and increase load factors by limiting airline departures. This authority was exercised in a different context in August 1981 when, in response to the strike by air traffic controllers, the FAA limited flight operations. The limitation effectively increased airline load factors and reduced petroleum consumption.

The greatest impediment to implementation of a rationing program is the problem of selecting a method of distributing rationing tickets that will be perceived as fair and that will work. As noted above, perceptions of fairness differ even between the members of the House of Representatives and the Senate, and thus there has been little success in finding a plan acceptable to all. However, even if a fair proposal for distributing rationing tickets to all consumers could be found, the physical problem of distributing the tickets is by no means simple. Indeed, one proposal offered by DOE would have cost more than $2 billion, required more than six months to start, and then would have required the voluntary participation of U.S. banks as distribution points. The bankers found the proposal sufficiently onerous as to threaten the very financial stability of the country.

The greatest problem with rationing, however, is neither the fairness by which ration coupons are distributed nor the procedure by which the tickets are delivered to consumers: It is the time required to put rationing into effect. Most studies suggest that at least six months would be required to put a rationing program into effect under the best of circumstances. This would mean that a disruption would run at least six months before emergency conservation through rationing took effect. Assuming that reductions in consumption were not realized by other measures in the meantime, much of the benefit of rationing would be lost.

There are other problems with rationing as well. Indeed, as Jacobs (1981) suggests, rationing may not have the intended effect of eliminating lines at filling stations if a rationing program is accompanied by price and allocation controls.

## Summary

Emergency conservation measures offer one viable mechanism for reducing consumption and mitigating the impact of a disruption. Roughly 2 million barrels a day of U.S. consumption could be cut by imposing a number of measures ranging from the innocuous to the draconian. However, as illustrated on Table 10-5, the source of the savings is not constant. Measures that would have offered substantial savings in 1973 (such as temperature controls) now appear to offer little potential because the benefits have already been realized. On the other hand, some measures that offered little benefit in the past

Table 10-5. Estimates of Emergency Conservation Potential.

| | White House Press Office (1973) | U.S. Department of Energy (1979) | National Petroleum Council (1981) |
|---|---|---|---|
| | | Thousand Barrels a Day | |
| Building temperature controls | 490 | 150-300 | 165 |
| Fuel switching to gas | 100 | 250-400 | 500-960 |
| Fuel switching to coal | 250 | 0 | 80 |
| Powerwheeling | – | 100-200 | 50 |
| Personal vehicle use | 1,150 | 340-635 | 1,235 |
| Totals | 1,190 | 840-1,535 | 2,030-2,490 |

(such as natural gas substitution) now offer large potential due to changes in the market.

This review of the various actions that have been suggested in the past leads to the following conclusions. First, the automobile and personal vehicle use are overrated as sources of conservation. Every study lists a number of small adjustments in vehicle use that, when added up, yield large savings. However, in practice, the measures are politically unpopular with one group or another and, if they are implemented, probably yield small savings.

Second, large reductions in fuel use can be achieved by increasing the use of natural gas in industrial plants and utilities. Indeed, the potential for oil savings by using gas is limited only by the capacity of industrial users and utilities to substitute gas for oil.

Third, while utility wheeling offers only limited savings because it is already taking place, large potential savings may be available if an aggressive program of electricity conservation is combined with a program of increased wheeling from coal based utilities to oil based utilities. The experience in Los Angeles during the 1973/74 disruption suggests that quick reductions in electricity use are achievable. Thus, the success of such a program would depend only on the effect of emergency conservation upon the capacity of power pools to wheel additional electricity. If wheeling capacity is increased when electricity use is cut, then electricity conservation offers significant potential as an emergency conservation measure.

## THE EFFECT OF IMPLEMENTING EMERGENCY CONSERVATION MEASURES AS A MEANS OF MODERATING THE EFFECT OF DISRUPTIONS

The primary purpose for which emergency conservation measures have been created is to reduce consumption quickly in order to prevent the imbalance between supply and demand, created by a disruption, from starting an upward spiral in the price of oil. This section assesses the effectiveness of such measures in stabilizing prices. One purpose of the analysis is to quantify the importance of quick action in cutting consumption. A second is to study the relationship between the size of a cut in consumption and the speed with which the cut can be imposed. Thus, this discussion will attempt to determine whether small cuts imposed immediately are more or less successful than large cuts imposed with a lag.

The analysis is performed with a stylized model of oil market behavior. A moderate-size disruption of 4 million barrels a day lasting at least twelve months is assumed to occur. It is assumed that only one action is taken to meet this disruption and that the action is taken by all consuming countries collectively. Four types of actions are assessed:

1. An immediate cut in consumption equal to 1 million barrels a day;
2. An immediate cut in consumption equal to 2 million barrels a day (half the loss of supply);
3. An immediate cut in consumption equal to 4 million barrels a day; and
4. A cut in consumption equal to 4 million barrels a day imposed four months after the start of the disruption.

The last of the cuts might represent a gasoline rationing program, which, as noted above, would require a long start up period, while the immediate cut in consumption would be analogous to bans on Sunday driving or on the sale of gasoline on weekends.

The effectiveness of each of these measures in moderating increases in spot values of crude due to the 4 million barrel a day disruption can be observed from Table 10-6, which shows the month-to-month change in spot values of crude on commodity markets.

The following conclusions can be drawn from these simulations. First, a cut in consumption will reduce the rate of increase in spot values. Second, an earlier cut in consumption has a more substantial impact on prices than a larger delayed reduction. Third, none of the cuts in consumption can stop the process of price escalation once it has begun, unless the reduction in consumption is greater than the loss in supply.

The failure of emergency conservation measures to stop increases in spot values is explained by the nature of the inventory acquisition process. The initial increase in prices on commodity markets (spot values) triggers inventory acquisition, which increases the size of the shortage. Thus, to stop the increase in prices, emergency conservation measures must not just cut consumption but must cut consumption below the level of supply. To illustrate this, consider the initial effect of a 4 million barrel a day decline in the supply of oil. The reduction creates a shortage of not 4 million but 5 million barrels a day, because the increase in spot values immediately increases the demand for inventories by 1 million barrels a day. Thus, emergency conservation measures that reduce consumption by 4 million barrels a day leave a shortage of 1 million barrels a day, which continues to raise prices. We conclude, then, that emergency conservation measures may be effective in fully meeting a disruption only if they overcompensate for the magnitude of the cut in supply.

It is also important to note that the size of the emergency conservation measure required to meet a disruption of a given magnitude increases with delay. This is illustrated on Table 10-7, which shows the amount of emergency conservation required to fully arrest a loss of 4 million barrels a day in supply if the action were delayed by one, two, or three months. The results indicate that, while a 4 million barrel a day cut would be adequate if it were made on the first day of the disruption, a cut in consumption of 5.7 million barrels a day would be required if consuming countries waited thirty days to respond. If action were delayed for sixty days, allowing the crisis to build, the size of the emergency conservation measures required to

Table 10-6. Effects of Conservation Measures on Increases in Spot Crude Values with an Interruption in Supply.

| | | Spot Crude Values | | | |
| | | Emergency Conservation Measures | | | |
| Month | Loss in Supply | 1 Million Barrel a Day Cut Immediately | 2 Million Barrel a Day Cut Immediately | 4 Million Barrel a Day Cut Immediately | 4 Million Barrel a Day Cut Delayed 6 Months |
| --- | --- | --- | --- | --- | --- |
| | Million Barrels a Day | | Dollars per Barrel | | |
| 0 | 0 | $ 30.00 | $ 30.00 | $ 30.00 | $ 30.00 |
| 1 | -4 | 37.27 | 37.27 | 37.27 | 37.27 |
| 2 | -4 | 46.80 | 44.44 | 39.90 | 49.23 |
| 3 | -4 | 68.26 | 55.64 | 43.15 | 70.80 |
| 4 | -4 | 93.51 | 74.71 | 47.25 | 109.70 |
| 5 | -4 | 137.43 | 104.94 | 52.56 | 168.53 |
| 6 | -4 | 199.58 | 146.08 | 59.24 | 210.43 |
| 7 | -4 | 284.64 | 200.58 | 69.99 | 257.51 |
| 8 | -4 | 395.82 | 270.17 | 86.02 | 307.19 |
| 9 | -4 | 532.63 | 354.77 | 107.23 | 355.45 |
| 10 | -4 | 688.50 | 451.37 | 132.13 | 397.29 |
| 11 | -4 | 849.23 | 553.11 | 160.44 | 427.50 |
| 12 | -4 | 993.98 | 649.42 | 191.33 | 441.80 |

Table 10-7. Size of Emergency Conservation Measures Required to Offset a Disruption of 4 Million Barrels a Day.

| Delay in Enacting Cut | Required Size of Cut in Consumption |
| --- | --- |
| Immediate response | 4 million barrels a day |
| Response occurs one month after disruption starts | 5.3 million barrels a day |
| Response occurs two months after disruption starts | 8.3 million barrels a day |
| Response occurs three months after disruption starts | 13 million barrels a day |

stop the price spiral increases to 8.3 million barrels a day because a panic demand for inventories has developed.

While these results overstate the magnitude of the measures required to bring a crisis to a halt, they emphasize again the point made in Chapter 8 that emergency measures must initially overrespond to events to bring the crisis to a halt. It is instructive to note that most past and currently contemplated emergency conservation measures include a component labeled "voluntary conservation" and, as such, represent underkill not overkill.

### Joint Tax and Conservation Measures

Emergency conservation measures offer limited benefits as measures for dealing with a disruption when they are imposed in isolation because, to be effective, the measures must be imposed very quickly and must compensate for both the loss in supply and the increase in inventory demand. This negative conclusion does not imply that these measures are of no use at all, but that the effectiveness of emergency conservation measures can be greatly improved if they are coupled with measures that both encourage liquidation of inventories and encourage consumers to reduce consumption through increases in prices. One such combined policy would consist of a tariff on oil imports plus conservation measures.

The effectiveness of joint policies can be shown by repeating the simulations of variable tariff programs discussed earlier. (Recall that

**Table 10-8.** Effect of Joint Tariff Conservation Measures as Mechanisms for Offsetting a 4 Million Barrel a Day Loss in Supply.

| | | Measures to Stop Increases in Spot Values by Year End | | Measures to Restore Spot Values Immediately | |
|---|---|---|---|---|---|
| Month | Loss in Supply | Tariff Only | Tariff Plus 2 Million Barrels a Day Conservation | Tariff Only | Tariff Plus 2 Million Barrels a Day Conservation |
| | Million Barrels a Day | | Tariff Required, Dollars per Barrel | | |
| 0 | 0 | $ 0.00 | $ 0.00 | $ 0.00 | $ 0.00 |
| 1 | -4 | 0.00 | 0.00 | 0.00 | 0.00 |
| 2 | -4 | 31.00 | 17.50 | 51.00 | 40.00 |
| 3 | -4 | 30.00 | 16.50 | 22.00 | 11.00 |
| 4 | -4 | 29.50 | 15.50 | 23.00 | 11.00 |
| 5 | -4 | 28.75 | 14.75 | 23.00 | 11.00 |
| 6 | -4 | 28.00 | 13.75 | 23.00 | 10.75 |
| 7 | -4 | 27.25 | 12.75 | 23.00 | 10.75 |
| 8 | -4 | 26.25 | 12.00 | 23.00 | 10.75 |
| 9 | -4 | 25.00 | 11.00 | 23.00 | 10.50 |
| 10 | -4 | 24.25 | 10.25 | 23.00 | 10.50 |
| 11 | -4 | 27.50 | 9.50 | 23.00 | 10.50 |
| 12 | -4 | 22.75 | 8.75 | 23.00 | 10.25 |
| Volume of 12 month stock drawdown—million barrels | | 687 | 160 | 764 | 376 |

the discussion in Chapter 8 focused on the size of tariff required to defend a particular price.) In the first simulation it was assumed that energy policymakers elect to act to prevent further increases in spot prices after the first month of the disruption, while in the second simulation it was assumed that policymakers act to defend the original price of oil. The results show that the size of the tariff required to achieve either result is substantially smaller than in the original simulation. It was determined in Chapter 8 that in the absence of any conservation measures, a $51.00 a barrel tariff was required to immediately restore prices to the predisruption level and a permanent tariff of $22.50 was required to maintain stable prices after the first month. When conservation measures are added, the initial tariff required during the first month is reduced to $41.00 a barrel, while the tariff required in the ensuing months is reduced to $10.00.

The use of emergency conservation measures together with the tariff also reduces the volume of oil required to defend a particular price level. For instance, simulations presented in Chapter 8 indicated that approximately 13 percent of precrisis stocks would be needed during the first year to defend a particular price level in the case of a 4 million barrel a day interruption in supply. However, when emergency conservation measures that reduce consumption by 2 million barrels a day are combined with the tariff, the stock requirement is reduced from 13 to 7 percent of precrisis stocks. The results of the policy simulations are given on Table 10-8, which indicates the tariff and rates of inventory liquidation required to defend two different price levels with and without emergency conservation measures.

## CONCLUSIONS

Emergency conservation measures offer a means of reducing the effect of an interruption in oil supplies. However, standby disruption plans should not consist of conservation programs alone, because the measures cannot be prepared in advance and will not offset the economic incentives to build stocks during a disruption. Instead, emergency conservation measures should be coupled with tax measures that will induce a reduction in consumption and measures which encourage the development and use of speculative stocks.

# 11 POLICIES FOR MANAGING DISRUPTIONS

Disruptions in world oil markets represent a serious threat to economic stability because they start a process that may lead to substantial increases in world oil prices—increases that will be maintained by anticompetitive behavior of certain cartel members and that over time will cause very large transfers of wealth from consuming countries to producing countries. The logical purpose of emergency energy policy measures should be to quickly eliminate excess demand on oil markets and thus stop the process of price escalation. To be effective, such policies must respond to the principal characteristics of the world oil market:

1. Crude oil prices set by OPEC move very slowly even when conditions of supply and demand move quickly.
2. Prices charged to consumers by oil companies also move slowly and tend to follow changes in OPEC prices.
3. Inventory levels and stock acquisition policies respond to differences between official crude oil prices and spot prices, creating a situation where inventories are acquired during times of shortage and liquidated during times of surplus.
4. Consumer demand adjusts very slowly to price changes.

Due to these characteristics, a loss in supply that theoretically should induce only a modest change in long-run oil prices could easily cause very large increases.

The analysis presented here has shown that these increases in prices may be blocked most easily if the disruption is met by bold policy measures that force consumer prices up very quickly while encouraging sales from inventories during the initial months of the crisis. At the same time, it was noted that the emergency measures had to be accompanied by other programs that would maintain the level of spendable income. These criteria, then, lead to the following conclusions concerning emergency measures:

1. The measures imposed for meeting a crisis should encourage immediate drawdown of both public and private inventories while at the same time inducing a maximum conservation response by consumers.
2. Since inventory drawdown is important during disruptions, measures to encourage stock building during times of surplus are essential.
3. Government reserves should be held in a manner that would facilitate immediate drawdown in the event of a crisis because early use of stocks has a more beneficial impact on world oil prices and also minimizes the use of stocks.
4. The tax receipts raised by imposition of the disruption tariff should be offset by recycling the increased tax revenues through standby rebate measures.
5. The optimal policies for dealing with a disruption consist of (a) a standby tariff on imports of crude and products that is initially imposed at a very high level and then gradually scaled down; (b) a mechanism for recycling tariff revenues; and (c) a program of emergency conservation measures that can be imposed quickly at the start of the disruption. Furthermore, while the list of emergency conservation measures must be prepared in advance, it must be updated continually as circumstances change.

## ELEMENTS OF AN EFFECTIVE POLICY

### 1. Measures for Meeting a Crisis Should Encourage an Immediate Drawdown of Stocks.

Every study of emergency preparedness emphasizes the importance of oil stocks but few address the issue of stock management during an interruption. As noted in the simulations presented in Chapters 6, 8, and 10, however, the drawdown of inventories represents one of the best, if not the best, means of moderating price increases during a disruption. This drawdown can be accomplished either through market mechanisms or by regulation. So long as the stocks are used, the rate of increase in prices on commodity markets will be slowed.

From an economic standpoint, the optimal means of encouraging stock drawdown would be by market mechanisms, specifically, by allowing those who speculate to make windfall profits. It is also tempting to argue that the volume of stocks liquidated will be greater (and thus the impact on the spot market greater) if market forces were used. (In particular, a tariff would both encourage sales from stocks and discourage consumption, while regulation-induced sales of stocks should not affect the demand for oil.) However, since these policies are untried, the evidence to support this assertion is purely theoretical.

### 2. Measures that Encourage Stock Building During Periods of Stability Will Pay Large Dividends During Periods of Disruption.

A principal conclusion of this study is that the use of stocks during a disruption can greatly reduce the rate of increase in commodity prices and crude oil prices. This effect is more easily obtained if stocks are built during periods of surplus. The analysis in Chapter 9 indicated that the incentive to increase stocks (or to hold speculative stocks) could be achieved either by increasing the expected return for holding stocks or by decreasing the cost of holding stocks.

Policy measures that would increase the expected return for holding stocks include explicit disavowal of price controls as a means for dealing with a crisis and expression of intentions to impose tariffs on imports of crude and products during a disruption. Both measures increase the return for holding speculative stocks.

Price controls and refiner taxes decrease the expected return for holding speculative stocks and would presumably decrease stock holding. Standby consumer taxes (or refiner taxes) also discourage speculative stock holding by reducing the return for holding inventories, while standby tariffs encourage speculative stock holding.

Measures that would reduce the cost of holding speculative stocks include interest subsidies, special tax treatment for windfall gains from holding inventories, and adoption of price supports. The measures that probably would be the most effective would be the payment of interest subsidies, treatment of the cost of storage as a tax credit, or a price support mechanism similar to that used in agriculture. Measures that delay the benefit until the date of the disruption (for instance, treating the gain on speculative stocks as a capital gain rather than ordinary income) would probably induce less stock building because the benefit would be uncertain and could even be rescinded in the future.

Some measures adopted during past disruptions tend to discourage stockbuilding and thus reduce preparedness. The principal villains here are price controls and volumetric allocations. Volumetric allocations are particularly troublesome: They reduce the incentive of firms receiving allocations to build stocks, since the cost of storing speculative stocks is a drain on corporate cash flow that is not incurred when oil is purchased under the allocation program. The same analogy applies to consumers entitled to fixed allocations.

Allocation programs also discourage stock holding by selling firms, because in the past they have been denied any speculative profits on the oil sold under the allocation program.

Finally, it should be noted that speculative stockbuilding may well be transitory. Stock levels may be increased during those periods of surplus when oil companies and oil consumers perceive that it is unlikely that price or allocation controls would be employed during a disruption and allowed to decline during those periods of surplus when companies and consumers perceive that the political climate makes it likely that price controls, allocations, and other forms of jawboning would be employed in the event of a disruption. Such behavior should be anticipated because expectations as to the likelihood of imposition of controls will affect the calculation of expected returns from speculation.

There appears to be no way to offset this impact of political conditions on stockholding. A statement by the President or the Secretary of Energy that controls would not be used will probably influence firms to build greater speculative stocks than they might build without the statement. However, actions in other regulatory areas can quickly offset the effect of any promise. Thus, speculative stocks would be expected to be built and held only if the actions of the federal government in other regulatory matters were fully consistent with the stated intentions of the president.

## 3. Government Emergency Reserves Should Be Held in a Manner that Would Facilitate Immediate Drawdown in the Event of a Crisis.

Many politicians, experts, and economists have endorsed building large governmentally owned emergency stockpiles. This advocacy is clearly correct because inventories of petroleum can be used to moderate increases in petroleum prices during interruptions of supply. However, this is true only if those stockpiles are used during disruptions. Furthermore, the greatest benefit per barrel (biggest bang for the buck) comes if those stockpiles are used early in a disruption in conjunction with other measures designed to cut consumption and trigger liquidation of private stocks. This means that programs for drawing down stockpiles must be developed at the same time oil is put into the reserve. This consideration has been totally ignored or given very limited attention in most studies.

Two studies that have addressed the issue of stock drawdowns are those by Alm, Colglazier, and Kates-Garnick (1981) and by Deese and Nye (1981). Alm, Colglazier, and Kates-Garnick (1981:1382) argue, "To minimize price hikes, oil must be released quickly and predictably at the early stages of an interruption." This view is fine as far as it goes, but they offer no quantitative rule. Deese and Nye (1981:403) are more specific, arguing, "To discourage panic stockpiling, the government should amend the Strategic Petroleum Reserve Plan [SPRP] with a schedule of stockpile releases that would be triggered automatically by supply interruptions of various magnitudes. Oil withdrawn from the reserve should be sold to the highest bidder at market prices." Unfortunately, this proposal does not go far enough either, because it is difficult to ascertain the magnitude of a supply interruption until it is well under way.

To facilitate the quick withdrawl of oil from the private reserve, an alternative proposal can be suggested under which shares (or calls) on oil in the reserve would be marketed to all interested parties. These instruments would entitle the holders to oil delivered upon demand for the period of duration of the instrument.

Under this proposal, the government (DOE) would purchase oil for the Strategic Petroleum Reserve, (SPR) using money borrowed on the nation's capital markets in the traditional fashion (i.e., increased government debt). DOE would then auction options or calls on the oil in the reserve in the same way the U.S. Treasury sells Treasury bills. These options would:

1. Entitle the holder to purchase oil from the reserve at a striking price for a fixed period of time, say, three months;
2. Have a striking price above the present market price of oil by some percentage—a percentage high enough to make SPR oil uneconomical except at the time of a disruption; and
3. Specify a delivery location, quality of crude, and delivery lag from the date of exercise.

This type of proposal would alleviate the need for enactment of a standby crude oil allocation program because crude-short refiners or product-short consumers could hedge. It would also relieve the government of the responsibility of deciding when to use the reserve. This is important because the federal government's performance in buying and selling commodities in the past has been abysmal, and because the SPR can be effective in reducing the impacts of a disruption only to the extent that it is drawn down during the disruption.

The adoption of a proposal of this sort would also allow the central government to subsidize the cost of storage explicitly if it elected to do so. This subsidy could be paid by accepting bids on calls that failed to cover the full interest cost of carrying the oil. That is, if 500 million barrels of storage were demanded at the market rate of interest (that is, if the government could sell calls on only 500 million barrels of oil if it insisted on fully covering the interest cost of the oil) but the government elected to hold more oil, it could elect to accept bids for more calls (say, a total of 700 million barrels) at a lower price while covering part of the cost of the oil out of general revenues. The most important advantages of this proposal, however, are, first, that the use of a financial instrument like a call would trigger immediate withdrawals of oil from the reserve at times of crisis and, second, that imposition of disruption tariffs would immediately trigger automatic withdrawals from the reserve.[1]

---

1. Some may be concerned that the use of market mechanisms to sell oil from the reserve would endanger the national defense, while others may worry that the use of the market endangers the competitive position of some smaller oil firms. The defense issue does create some problems. Indeed, it is certainly appropriate for the federal government to hold separate defense stocks, presumably paid for out of the defense budget.

The competitive issue should cause no problem. Firms in the oil business should recognize the risks of a disruption and presumably can make the preparations that best suit their

## 4. Measures to Recycle Increased Tax Receipts Should Be Prepared in Advance of the Disruption.

The economic losses that accompany a disruption in oil markets result from the transfer of income from consuming countries to producing countries during and after the disruption. These income transfers are caused by the increase in the price of oil. To the extent that increases in oil prices can be prevented, it is possible, then to reduce the macroeconomic damage caused by the disruption. However, quick recycling of increased receipts from tariffs or taxes will be required in order to minimize that damage.

The CBO (1978) study, *Direct Federal Action on Oil Imports: An Analysis of Import Fees and Quotas* nicely demonstrates this phenomenon. CBO found that the imposition of a $5.00 a barrel tariff on imported oil in the late fall of 1978 (at a time when the average cost of imports was estimated by CBO to be $15.00) would have raised the 1979 unemployment rate from 6 percent to 6.2 percent and raised the predicted 1980 unemployment rate from 6 percent to 6.4 percent. CBO also found that the tariff would cut the level of 1979 real GNP by 0.6 percent and 1980 real GNP by 1.2 percent. These findings assumed that receipts from the tariff would not be recycled to the economy. The CBO (1978:20) study also concluded that there would be no adverse economic effects if the same tariff were coupled with a program that fully refunded the tariff receipts. This conclusion is in accordance with most other classic economic studies of the effects of taxation, which show that taxes have little or no impact on the level of real GNP or the rate of unemployment as long as the increased receipts are immediately recycled.[2]

---

needs. If they face shortages of oil in a disruption, they could, under this proposal, purchase calls on the reserve. Alternatively, they could find other mechanisms, including accepting the risk of the loss in supply. The point is that the federal government should not be required to bail them out in an emergency.

2. It must be noted, however, that these "classic" findings are being challenged by many economists who argue that changes in tax rates change incentives at the margin and thus have a macroeconomic effect. See, for instance, White House Press Office (1981:IV.2), where this potential effect is acknowledged. However, it can be argued that a tax change that did not change any marginal tax rate but that increased revenues would have no macroeconomic effect if equal compensating changes were made so that the increased revenues were immediately recycled.

Unfortunately, the development of an adequate recycling mechanism is very difficult, given the size of the tariff that might have to be imposed and the structure of the U.S. tax system. In the case of a severe disruption, it might be necessary to impose a fee of as much as $100 a barrel, which would then be scaled down over a period of ten to twelve months, so that the average fee during the year might be as high as $40 to $50 a barrel. In such circumstances the receipts from the fee plus receipts from increased corporate profits taxes and windfall profits tax might easily exceed $200 to $300 billion dollars—an amount roughly comparable to federal receipts from individual or corporate tax payers. If not recycled quickly, the fiscal drag from these tariffs could promptly create a recession.

Under the U.S. system of income taxation, recycling of these receipts is not a simple task. Because the federal income tax system is very clumsy and lethargic, a number of weeks would be required to make payments or reduce tax withholdings, and some individuals would not initially receive any benefits.

We are left with a troubling dilemma. On the one hand, the costs of not responding are quite high; on the other hand, a quick response without quick recycling of the transferred income may create a major recession. To surmount this problem, legislation should be passed permitting the federal government to immediately apply rebates of specified amounts as soon as fees are imposed. This would permit the Treasury Department to immediately increase Social Security and other payments by a specific amount for every $1.00 of fee imposed. Having such a program prepared in advance would reduce the recycling delay.[3]

---

3. The transitional difficulties could be handled more easily if the United States were to adopt a value added tax in lieu of the federal income tax. These taxes are simply a form of sales tax. If such a system were in place (as they are in most European countries), then revenues from the emergency tariff could be easily recycled by making compensating reductions in the tax rate on other goods. This is not to imply that the threat of an interruption in petroleum supply is evidence by itself for the adoption of a value added tax, but only that there is ample evidence that a value added tax could be used as a means for discouraging consumption and encouraging both savings and investment and that the use of a value added tax instead of an income tax would reduce the risk of a recession resulting from the imposition of a disruption tariff.

## 5. An Effective Disruption Policy Should Include (1) a Standby Tariff, (2) Emergency Conservation Measures, (3) Private Control over SPR Oil, and (4) a Mechanism for Recycling the Increased Receipts.

Throughout this study it has been argued that interruptions in oil supply are transformed from minor interruptions to major catastrophes because markets adjust slowly. The speed of adjustment in OPEC prices is so slow that ten to twelve quarters are required to recapture the full amount of any change in spot values. Consumer prices demonstrate an identical slow rate of adjustment, while the adjustment in final demand to the eventual change in consumer prices is even slower—requiring perhaps four years. Finally, the process is complicated by the perverse direction of adjustment in the level of inventories.

Given these characteristics, it seems obvious that an effective policy would have to speed the process of adjustment and, in the case of inventory behavior, reverse the natural response. To achieve such a change, a four-step disruption program is recommended:

1. Large temporary disruption tariffs should be used to quickly cut consumption and encourage sales from stocks;
2. Simple and easily administered emergency conservation measures should be imposed quickly at the start of an interruption;
3. Private control should be adopted over the distribution of oil held in the strategic petroleum reserve; and
4. Programs should be perpared that would permit quick recycling of revenues from the tariff.

In addition, the adoption of tax or subsidy programs is recommended to encourage firms to build inventories during periods of stable supply.

This program would encourage the development of stocks during periods of stability, make it profitable for the holders of these stocks to release them at the start of a crisis, and encourage consumers to rapidly cut consumption and switch to other fuels. At the same time, the adoption of emergency programs to recycle receipts from the taxes would be required to prevent these measures from doing irreparable harm to the economy.

The key element in the program is the graduated disruption tariff. It should be very large initially but be scaled back gradually as consumers adjust to the changed conditions of supply. The graduated nature of the tariff serves three purposes: to shock consumers, to encourage consumers to postpone purchases, and to encourage those holding speculative stocks to sell them quickly. The large initial value of the tariff will induce a very sharp reduction in consumption. Furthermore, the promise of declining prices due to the prospective adjustments in the tariff should cause consumers to postpone consumption wherever possible. Both actions will reduce upward pressure on spot prices by reducing demand. At the same time, the prospect of higher current than future prices should induce holders of speculative stocks to sell early, further reducing upward pressures on spot prices.

The second element in the program should be emergency conservation measures. The measures used should be those that can be imposed quickly and can be easily administered. Among the measures that probably would achieve quick reductions in consumption and could be easily administered are:

- Mandated reductions in electricity use to permit increased wheeling of electricity;[4]
- Closing of gasoline stations on weekends and perhaps a ban on driving for one or two days during the weekend; and
- Coal and natural gas substitution to the extent that substitutes can be made quickly and are available.

By and large, emphasis should be placed on the visible but potentially unpopular measures at the start of a disruption rather than the more popular, less visible ones that have been put forward during previous crises. The less visible measures generally do not offer the same reductions in oil use, either because people resist using them or because they are already being used. Building temperature controls offer a perfect illustration. In 1973 and 1979 emergency energy planners projected large savings from changes in building temperatures. However, a recent NPC study found that only minimal potential sav-

---

4. This measure will work adequately today but perhaps not as well in the future as utilities gradually move away from oil. Obviously, for this to work, some utilities must still be using oil. An equally necessary but less obvious requirement is that refiners have the flexibility to turn the oil saved into other products.

ings from thermostat adjustments exist because the adjustment has already been made.

Third, the program should permit private parties to control distribution of oil held in the strategic reserve. This control could be achieved either by offering supplies at a predetermined price that would be unattractive under normal circumstances but that would become attractive if, as, and when a disruption occurred, or by creating some type of financial instrument such as a call or convertible bond that could be marketed to the public and that could be exchanged for oil. The adoption of either proposal would allow the oil held in the reserve to be used quickly to decrease the upward pressures on prices that occur during a crisis.

The argument for private control over public oil reserves is basically the same as the argument for use of a disruption tariff: speed of response. The oil in the reserve is more likely to be used early in a disruption if control is left in private hands, than if it is left in the hands of public officials. This will mean that the oil in the reserve will have a more favorable effect on prices.

It should also be noted that the adoption of this policy would not, as some suggest, leave consumers vulnerable to prolonged disruption, *if* a disruption tariff designed to bring supply and demand back to equilibrium were coupled with the stock drawdown. In fact, as the calculations presented in Chapter 8 suggest, fewer stocks are required to meet a disruption when the stock drawdown occurs quickly. The worry that stocks will be used up is not, and should not be, the concern of energy planners—assuming, of course, that there is no attempt to prevent consumer prices from increasing.

Finally, the program should encourage private parties (consumers and refiners) to build stocks. What should be the size of the incentive? The answer is, unfortunately, not clear, because the data on the costs and demand for storage are not readily available. Thus, more research and possibly some experimentation will be required to determine the proper subsidy.

## THE INTERNATIONAL ENERGY AGENCY AS A POLICY INSTRUMENT

The International Energy Agency (IEA) is the organization that would coordinate consumer country response to a disruption. At present, planned IEA response to an interruption in supply (the International Energy Program or IEP) consists of sharing available supplies of oil, coordinating emergency conservation plans in the various consuming countries, and coordinating the drawdowns of emergency stockpiles. These programs would be triggered only in the event of a shortfall in supply greater than 7 percent of projected supplies for all member countries, or in the event that one member country experienced a loss of 7 percent of its supply. The IEP envisions a two-stage response. If the loss of supplies is between 7 and 12 percent of projected supplies (approximately 2.5 to 3 million barrels a day), the loss is to be made up by emergency conservation measures. If the loss in supply exceeds 12 percent, it will be met by a combination of emergency conservation measures and stock drawdown (Deese and Nye 1981:412).

If some or all consuming countries experience a shortage in excess of 7 percent, the IEA sharing plan requires all member nations to share their supplies with other nations. The plan would allocate available oil supplies to various countries according to historical rates of consumption less "demand restraint" imposed by the IEA, less a country's "emergency oil drawdown obligation" (ERDO). This last item represents a country's allowable inventory drawdown. It can, however, be met by additional demand restraint, substitution of other forms of energy, or emergency production of oil. If a member country's available supply including imports exceeds its allocation right as computed by this formula, it would be required to sell oil to other IEA countries; if its available supply falls short of the allocation right, it would be permitted to purchase oil.

To illustrate this computation, calculations of supply rights in the case of two disruptions for the IEA as a whole and for the United States are reproduced from the NPC study (see Table 11-1).

In Scenario 1, the NPC assumes that supplies available to the IEA nations fall 5.1 million barrels a day below projected demand (a 13 percent shortfall). The International Energy Program would require member countries to institute mandatory measures to reduce consumption by 10 percent, or 3.7 million barrels a day. This would leave a remaining shortage of 1.2 million barrels a day to be met by emergency oil drawdown obligations on the individual countries. Individual country ERDO obligations would be allocated according to the country's individual share of base period consumption. According to the NPC, the U.S. obligation would be 38 percent.

Table 11-1. Supply and ERDO Calculations for Two Disruptions.

|  | Scenario 1<br>Loss of 5.1 Million<br>bbl/day of Supply | Scenario 2<br>Loss of 7.8 Million<br>bbl/day of Supply |
|---|---|---|
| *IEA as a Whole* | Million Barrels a Day | |
| Available supplies | 32.4 | 29.5 |
| Permissible consumption | | |
| Base period consumption | 37.3 | 37.3 |
| Less 10% demand restraint | 3.7 | 3.7 |
| Net Consumption | 33.6 | 33.6 |
| Group supply shortfall | | |
| Permissible consumption | 33.6 | 33.6 |
| Less available supplies | 32.4 | 29.5 |
| Shortfall (ERDO) | 1.2 | 4.1 |
| *United States* | | |
| ERDO (38% of group supply shortfall) | 0.5 | 1.5 |
| Permissible consumption | | |
| Base period consumption | 17.5 | 17.5 |
| Less 10% demand restraint | 1.7 | 1.7 |
| Permissible consumption | 15.8 | 15.8 |
| Supply right | | |
| Permissible consumption | 15.8 | 15.8 |
| Less ERDO | 0.5 | 1.5 |
| Supply right | 15.3 | 14.3 |

Source: National Petroleum Council, 1981, *Emergency Preparedness for Interruption of Petroleum Supplies into the United States*, Washington, D.C., The National Petroleum Council, p. M-14.

The U.S. obligation as calculated by the NPC would be to cut consumption by 10 percent or 1.7 million barrels a day and, in the case of a 5.1 million barrel a day shortfall, to make emergency oil drawdowns in the amount of 500,000 barrels a day. In the case of a larger interruption (7.8 million barrels a day), the NPC calculations indicate that the IEA countries as a group would be required to make emergency drawdowns at a rate of 4.1 million barrels a day and that the U.S. share would be 1.5 million barrels a day. Based upon these calculations, the NPC shows that the U.S. would be entitled to a supply of oil (domestic production plus imports) of 15.3 million barrels a day in Scenario 1 and 14.3 million barrels a day in Scenario 2.

There have been a number of criticisms of this sharing scheme. Krapels (1980), for instance, suggests that the IEA should attempt to manage "sub-trigger" crises (interruptions in supply of less than 7 percent) and offers three reasons for this argument. First, IEA countries with large reserves must be willing to shoulder a greater burden for meeting a disruption. Second, those unlucky countries that happen to be dependent upon the disrupted supply of oil will have to draw down their inventories at a greater rate than those that, by luck, do not find their supplies disrupted. Third, "It is unlikely that 'consultations' between governments and companies will have a sufficiently uniform effect on supply distribution and stock drawdown or buildup. Any substantial effort to relieve the burden of countries which are suffering disproportionately large losses of oil involves too many millions of dollars and too many strong corporate national interests to be taken informally" (Krapels 1981:47). To remedy this problem, Krapels recommends that the IEA intervene early in even the smallest disruption to guarantee that supplies will be reallocated over time so that those countries suffering from a disruption will be willing to draw down stocks with the knowledge that they will later be resupplied.

Krapels fails to deal with the fundamental structural problems present in the design of the IEA. Three of these problems are identified by Deese and Nye (1981), who find that:

- The IEA is weakened by the presence of a major consumer, France;
- The IEA's ability to cope with a crisis is compromised by its inability to respond quickly to a crisis; and
- The IEA is further weakened by its unwillingness to take a confrontational position with OPEC.

Only the second of these criticisms is important, but it is sufficient by itself to render the IEA useless.

The problem with the IEA, or any organization designed to cope with an economic crisis, is that such organizations are, by their very nature, incapable of responding quickly. Instead of responding at the start of a crisis, they begin an elaborate process of meetings and consultations to ascertain the magnitude of the problem and discuss remedies. Inevitably the discussion of remedies becomes a process of bilateral negotiations, and, inevitably, the crisis worsens while the discussions take place.[5]

In the case of oil markets, it can be anticipated that the discussion among the IEA ministers would take place during the first month or two after an interruption began, while prices on spot markets were being pushed rapidly upward, increasing the long-run consequences. These price increases would set in motion the process of hoarding and stock building, which has been shown to worsen the crisis.

Even a quick decision by the IEA to act would not begin to resolve the problems created by an interruption in supply, because the present IEA emergency programs are neither adequate nor appropriate to the problem at hand. First, they address quantity, not price, and second, they are not strong enough to sort out the disorder created by the disruption.

The focus of the IEA emergency plans on quantity is understandable but unfortunate, because it presupposes that the members of the IEA and the IEA Secretariat will know the magnitude of the supply loss and be able to forcast demand before the shortage. Several critical assumptions are required to make this estimate. Two of the more important are (1) that the IEA countries can correctly ascertain the share of any shortfall the IEA member countries will have to absorb and (2) that the size of the shortage in the IEA will not increase as rising spot product values cause firms to increase inventories. While it may be possible to prorate the shortfall between IEA and non-IEA consumers to solve the first problem, it is probably not possible to curtail stock acquisition because inventory demand is positively related to differences in prices. Thus, in attempting to assess the size of the shortage, the IEA must recognize that it is chasing a moving target and that the gap between available supply and demand at prevailing prices will increase with time.

---

5. The record of the IEA in the Iranian crisis is interesting. The crisis began in November 1978. By December it was obvious that at least 2 million barrels a day in supply had been lost, and, according to *PIW*, the IEA was watching matters "closely." In the meantime, the situation worsened. In March, the IEA governing board finally acted, but by then the damage had been done.

The second problem with the present IEA emergency program is that it relies primarily on emergency conservation measures. These measures presumably will be prepared in advance by member countries. However, as noted above, emergency conservation measures are neither static nor easy to impose. Thus, in the event of an emergency it is easy to imagine the U.S. Secretary of Energy announcing a group of conservation measures designed months earlier, only to find that many of them have already been implemented by cost conscious consumers. It is even easier to imagine that political considerations would make it impossible for him to order the implementing of measures that really would work (such as weekend closing of gasoline stations or bans on Sunday driving).

The third problem with the IEA plans is that they represent "underkill" because they will probably make cuts in consumption that do not equal the loss in supply. Of course, the IEA plan anticipates this problem and would offset it by stock drawdowns (ERDOs). Here, however, one must anticipate a conflict between governmental and private incentives. The member governments will attempt to meet a shortfall by encouraging stock drawdown, while private interests (both consumers and oil companies), anticipating increases in prices, will attempt to stock up. The result will be chaotic. As Deese and Nye (1981:419) note, "Panic in even one of these major consuming nations could disrupt the entire market system." They council ". . . that any agreement to prevent such a panic will require creditable assurance that remaining supplies be shared fairly."

In short, the International Energy Agency plan focuses on the wrong target (quantity), is too small to meet the problem, and requires an excessive period of time to be implemented. It will not work.

## CONCLUSION

Throughout this volume it has been argued that policies for meeting disruptions in oil markets should emphasize speed of response and attempt to offset the natural incentive to hoard with temporary incentives to reduce stocks. It has frequently been suggested or implied that an analogy exists between disruptions on oil markets and financial panics. One does. Banking and oil market disruptions have many identical characteristics, and the solution to banking panics will, if properly applied to oil markets, yield the same beneficial results. In financial markets, the solution is to lend, and lend freely.

> The way in which the panic of 1825 was stopped by advancing money has been described in so broad and graphic a way that the passage has become classical. "We lent it," said Mr. Harmon on behalf of the Bank of England, "by every possible means and in modes we have never adopted before; we took in stock on security, we purchased Exchequer bills, we made advances on Exchequer bills, we not only discounted outright, but we made advances on the deposit of bills of exchange to an immense amount, in short by every possible means consistent with the safety of the bank, and we were not on some occasions over-nice." (Bagehot 1873: 51.)

There can be no better policy for meeting a disruption in world oil markets. The adoption of such policies, however, would require considerable courage on the part of public officials, politicians, and officials of the large oil companies. Unfortunately, during past crises, courage is a commodity that has been in even more limited supply than oil. It can only be hoped that during future disruptions the situation will be otherwise.

# APPENDIX
*Marshall Thomas*

## THE ABCs OF MEASURING OIL MARKET PRICE TRENDS

### Spot Prices—The Trend Setters

The volume of oil bought and sold on a spot market basis is relatively small (estimated at anywhere from 5% to 15% of total supply), but it provides a remarkably accurate indicator of long-term trends. Spot markets have signaled and sometimes even precipitated OPEC pricing actions, and they have often set the tone and direction of oil prices in mainstream consumer markets. The link is not always a fundamental economic tie; more often it simply reflects the reality that buyers and sellers, producers and consumers have not found a better indicator of the ever-shifting balance between demand and supply.

Many observers associate the spot market with the large Rotterdam entrepot, but the market is not physically located in any one place. Spot sales and purchases occur worldwide, whenever and wherever willing buyers and sellers make contact—whether in person, by telephone or telex. PIW focuses its spot market coverage on a sampling of the bid and asked prices of key crudes and products in selected locations.

Reprinted by permission of *Petroleum Intelligence Weekly*, copyright 1981 Petroleum and Energy Intelligence Weekly Inc.

### Spot Crude Oil Markets

Though some quantities of spot crude oil are bought and sold on a delivered basis to end-user markets such as Rotterdam, the U.S. Gulf Coast and Japan, most transactions are pegged to the f.o.b. prices at export loading ports in oil producing nations. Prices are commonly cited for spot (and contract) crude supplies, with most buyers making their own tanker transport arrangements. When suppliers have their own transport available they will sell on a delivered basis, but prices are still generally equivalent to the f.o.b. price plus freight. When unsold cargoes are put "on the water" en route to end-user destinations, sellers are sometimes forced into "distress" sales at well below the f.o.b. price plus freight.

Before the 1980 OPEC price eruption, spot market prices for crude oil were fairly well defined, and it was possible to measure market trends accurately within a few cents a barrel at any time. But with prices now scaling the $40 level, pricing tolerances have widened considerably. Spot prices can now vary by plus or minus 10¢ to 15¢ (a 20¢ to 30¢ swing), depending on the exact date of loading, vessel size, whether a full or partial cargo, and numerous other factors. Thus it is possible for Arabian Light crude to sell at $39.75 a barrel in one case and $39.50 in another—almost simultaneously. The sequence of the two transactions could lead to an erroneous conclusion that spot market levels are rising or falling, when in reality both may reflect the tolerances within the same market.

As a broad measure of trends, PIW traces the f.o.b. prices of three primary categories of export crude since specific grades are not all regularly bought and sold on a spot basis: (1) The Mideast 34-gravity crude category encompasses such grades as Arabian Light, Iranian Light, Basrah Light, and Dubai; (2) Mideast 31-gravity includes Arabian Medium, Kuwait and Iranian Heavy; and (3) African Light 37/44-gravity comprises such crudes as Algerian Saharan, Libyan Zueitina, Nigerian Bonny Light and Brass River, as well as North Sea Ekofisk. There is a growing tendency in oil industry practice to refer to the third group as "Cadillac" crudes—that is, the top quality, most expensive oils.

PIW elects to show prices by crude categories rather than mathematically averaging estimated prices for an exhaustive list of crudes.

## Spot Product Markets

Though spot supplies of refined products are bought and sold in many locations (including inland domestic markets in the U.S., Europe and Japan), PIW limits its coverage to six major refining centers. Five of them—Rotterdam, the Mediterranean (Italy), the Arabian/Persian Gulf, Singapore and the Caribbean—are primarily export oriented. The U.S. Gulf Coast is used as a reference point for the U.S. since it is a prime source of supply for the big U.S. Midwest and East Coast markets, as well as a major center for U.S. crude oil imports.

Price fluctuations on the spot product market occur on a day-to-day, and often minute-to-minute basis. This can make it difficult to perceive rhyme or reason for the fluctuations. Prices can spurt because one buyer comes to the market and doesn't quickly find a seller, or vice versa. As a result, market shifts on any given day may have little to do with the fundamental demand/supply.

Rather than trace the day-to-day swings, PIW focuses on typical prices for key products over a month-long period for a more reliable analysis of trends.

Few if any oil companies now secure crude oil for the sole purpose of refining it to sell 100% of the product yield on a spot basis. In most cases, refiners sell spot only when they have surpluses of a given product and buy to fill out their own shortages. In a sense, the spot market is serving as a supply balancing mechanism for refiners and marketers.

## Comparison of Spot Prices for Refined Products and Crude

The real spot market value of a producer's barrel of crude is determined in the spot markets for both crude and products. A valid price comparison of the two requires a series of calculations. The first step is to convert the range of various refined products into crude oil equivalents. This is done by determining the proportion of each major refined product (naphtha, gasoline, gas oil, fuel oil, etc.) that can be produced from a barrel of a given type of crude oil. The proportionate yield or mix of refined products varies not only with the quality of each crude oil, but also with the regional needs of each consuming area, with shifting winter and summer demand patterns and with the technical capabilities of individual refineries.

## Determining Refined Product Yield

PIW has developed data on crude oil yield patterns that are "typical" or "representative" of the refining industry in each of the six major refining centers. Yield patterns are based on refinery capabilities in handling the marginal or "last barrels" of crude oil supply instead of the "first barrels" that are ordinarily earmarked to meet each refiner's basic end-user demand requirements.

Yield patterns for the United States reflect a more sophisticated refining system, one designed to upgrade heavier products (fuel oil and distillates) into larger shares of lighter products (gasoline and heating oil). These more expensive systems allow refiners more flexibility in selecting crude oils. They also add value to many crudes, since light products are usually priced higher than heavier products.

The following table shows the basic differences in the refined product yield from a barrel of Saudi Arabian Light crude in a U.S. refinery and in a European plant—both on an upgraded and normal basis.

The export refining centers in Europe, East of Suez (in the Gulf and Singapore) and the Caribbean are still mainly geared to the sim-

Arabian Light Crude Oil 34-gravity (*in Volume Percent*)

| Product Type | U.S. Refinery | Europe Refinery | |
| --- | --- | --- | --- |
| | | Upgraded | Normal |
| Naphtha | — | 7.8% | 6.7% |
| Premium Gasoline | 6.9% | 12.7 | 7.7 |
| Unleaded Gasoline | 13.7 | — | — |
| Regular Gasoline | 25.1 | 9.9 | 5.8 |
| Gas Oil | 25.0 | 33.2 | 36.6 |
| Total Light Products | 70.7% | 63.6% | 56.8% |
| Fuel Oil | 25.7 | 32.6 | 38.5 |
| Refining Fuel and Loss | 3.6 | 3.8 | 4.7 |
| Total | 100.0% | 100.0% | 100.0% |

pler topping/reforming refinery units. For the present, therefore, PIW's yield basis reflects this prevailing pattern. The product supply mix will eventually change, however, as some existing refineries are upgraded with cracking facilities and as new plants start up in the Middle East and North Africa.

## Calculating the Value of the Refined Barrel

The first step is to compute the weighted average value of all the refined product components in a barrel of crude oil at the refinery gate. In trade jargon this is known as the "Gross Product Worth" or GPW. This is determined by multiplying the prevailing spot price for each product by its percentage share in the yield of the total barrel of crude oil.

The value of the fuel oil portion of the yield is largely determined by its sulfur level. Therefore an adjustment must be made where the sulfur level in fuel oil produced from a given crude differs from the prevailing quality of fuel sold in the spot product market. Generally, a refiner will blend fuels of various qualities to meet each market's needs.

Following is a sample calculation of the Gross Product Worth of a barrel of Arabian Light crude refined in Rotterdam with its product yield sold at spot market prices (assuming December 1980 prices).

### Arabian Light

| Product Type | Rotterdam Spot Price | | Product Yield | | Value Of Yield |
|---|---|---|---|---|---|
| Naphtha | $38.65 | x | 6.7% | = | $2.59 |
| Premium Gasoline | 43.28 | x | 7.7 | = | 3.33 |
| Regular Gasoline | 41.85 | x | 5.8 | = | 2.43 |
| Gas Oil | 40.10 | x | 36.6 | = | 14.68 |
| Fuel Oil: | | | | | |
| 1% Sulfur | 35.51 | | | | |
| 3.5% Sulfur | 33.71 | | | | |
| Adjusted Fuel Price* | 33.82 | x | 38.5 | = | 13.02 |
| Total Value of Arabian Light's Product Yield (GPW) | | | | | 36.05 |

*Adjusted to reflect value of 3.18% sulfur fuel oil yield of Arabian Light crude. PIW allows a 50% (or half-way) sulfur "blending" credit or debit where the sulfur level deviates from grades normally traded on the spot market.

## From the Refinery Gate Back to the Crude Oil Loading Port

The Gross Product Worth of a refined barrel is not strictly comparable with spot or official selling prices of a crude at the oil producer's loading port. To make it comparable, the costs of transporting crude to the refinery and of refining itself must be deducted. The marketing costs usually associated with sales to end-user consumers are not a factor, since most spot deals are wholesale in fairly large quantities at the refinery gate.

### The Costs of Crude Transport and Refining

Since spot product markets are mainly supplied by the "last barrels" of crude a refiner handles, transport and refining costs also are figured on a marginal or incremental basis.

For crude transport, it is the cost of chartering an appropriately sized tanker on the spot market for a single voyage. The transport cost yardstick is set by Worldscale, a trade association which publishes a base or "flat" rate for voyages between each oil loading and receiving port. Day-to-day tanker market fluctuations are measured in Worldscale "points," which are a percentage of the standard "flat" rate. The cost of chartering a ship at Worldscale 36 would be 36% of the flat rate for the spot cost of transport to Rotterdam:

*Flat Rate* Per Long Ton                                    $26.88
  *Per Barrel Arab Light                                    $3.59
*Spot Cost* for 200,000-Ton Tanker at Worldscale 36
  36% × $3.59 = $1.29 Per Barrel Cost

*Converted at 7.49 barrels per ton for 34-gravity crude.

Refining costs consist of the "out-of-pocket" operating expenses involved in the handling of the "last barrel" of crude by a refiner. This does not include the amortization and depreciation expenses usually charged against base-load operations.

PIW figures "out-of-pocket" costs at 65¢ a barrel in the U.S. and 30¢ elsewhere. A portion of the crude oil yield is used as refinery fuel and usually also, a small amount is lost in the refining process. Rather than place a value on those portions, PIW simply subtracts the volume from the finished product yield.

### At the Loading Port

The entire complex calculation now comes into focus. By subtracting freight and refining costs, downstream spot product prices are translated into an equivalent crude oil value at the loading port—the so-called "f.o.b. netback."

| | |
|---|---:|
| *Total Value of Arabian Light's Product Yield (GPW)* | $36.05 |
| Less: Incremental Refining Cost | -0.20 |
| Spot Freight Cost (At Worldscale 36) | -1.29 |
| *Implied F.O.B. Value Of Refined Arabian Light Crude* | 34.56 |

### Profits or Losses

The spot product market is a winner or loser depending on each refiner's crude oil supply costs. Generally, the spot value of the refined crude barrel at the producer's loading port is compared with either official or spot crude prices to determine profit or loss. In most cases, those refiners getting crude at straight official prices can make a profit; those buying at spot prices sustain losses.

| | |
|---|---:|
| *Implied F.O.B. Value of Refined Arabian Light Crude* | $34.56 |
| Versus Arabian Light Official Price | 32.00 |
| Refiners' Profit Margin | +2.56 |
| Versus Arabian Light Spot Price | 39.35 |
| Refiners' Profit Margin | -4.79 |

The price trends of recent years (page 17) suggest that spot product prices are reflecting refiners' averaged crude oil costs, including relatively lower cost crude at official prices.

For many oil buyers, it is now more economic to buy spot products instead of spot crude oil, removing much of the incentive for the traditional third-party processing arrangements common in the 1960s. Those deals involved buying cheap incremental supplies of spot crude, "renting" surplus processing capacity from a refiner, and selling the resulting refined product yield on spot markets like Rotterdam. The processing fee was set at a price above the refinery owner's "out-of-pocket" incremental operating costs.

### Measuring the Official Pricing Structure

After determining the absolute spot product value of major crude oils, PIW measures the value of each crude against Arabian Light, OPEC's traditional "marker" base. This shows whether the variations or "differentials" among the official crude prices are aligned with spot product market realities. The differences in the crude values reflect each oil's quality and the refiner's distance from the producer nation.

Spot Values of Major Crudes in Two Key Refining Areas in 1980 (dollars per barrel).

| U.S. Gulf Coast | Jan. | Feb. | March | April | May | June | July | Aug. | Sept. | Oct. | *Nov. | *Dec. |
|---|---|---|---|---|---|---|---|---|---|---|---|---|
| Arab Light–34 | 33.49 | 32.66 | 31.57 | 33.46 | 34.93 | 33.20 | 32.80 | 31.91 | 30.77 | 32.07 | 34.23 | 34.51 |
| Arab Heavy–27 | 28.00 | 26.65 | 24.61 | 25.47 | 28.63 | 27.53 | 28.29 | 28.01 | 26.27 | 28.36 | 31.14 | 31.39 |
| Iran Light–34 | 34.27 | 33.54 | 32.55 | 34.59 | 35.85 | 34.13 | 33.48 | 32.51 | 31.42 | 32.66 | 34.74 | 35.03 |
| Iran Heavy–31 | 33.13 | 32.29 | 31.12 | 32.95 | 34.53 | 32.93 | 32.52 | 31.69 | 30.47 | 31.87 | 34.07 | 34.33 |
| Kuwait–31 | 30.39 | 29.24 | 27.54 | 28.85 | 31.34 | 29.97 | 30.20 | 29.68 | 28.18 | 29.97 | 32.48 | 32.68 |
| Nigeria Light–37 | 37.86 | 37.11 | 36.41 | 38.93 | 39.32 | 37.41 | 36.45 | 35.16 | 34.59 | 35.22 | 37.20 | 37.52 |
| Algerian Saharan–44 | 37.13 | 36.77 | 35.91 | 38.66 | 39.16 | 37.07 | 35.79 | 34.86 | 34.79 | 34.38 | 36.65 | 36.55 |
| Libya Zueitina–41 | 37.32 | 36.94 | 36.03 | 38.30 | 39.15 | 37.10 | 35.80 | 34.93 | 34.59 | 34.61 | 37.35 | 36.61 |
| Venezuela Tia Juana–31 | 34.49 | 33.45 | 32.31 | 34.45 | 35.34 | 34.09 | 34.21 | 32.74 | 32.20 | 33.38 | 35.72 | 35.86 |
| Indonesia Minas–34 | 35.82 | 34.64 | 33.44 | 35.63 | 36.34 | 34.82 | 33.93 | 32.93 | 32.16 | 32.91 | 35.25 | 35.23 |
| North Sea Ekofisk–42 | 35.93 | 35.26 | 34.02 | 36.50 | 37.86 | 35.87 | 35.07 | 34.37 | 33.79 | 33.92 | 36.83 | 36.20 |
| *Rotterdam* | | | | | | | | | | | | |
| Arab Light–34 | 34.09 | 31.17 | 30.14 | 32.20 | 31.98 | 31.29 | 30.24 | 28.12 | 29.47 | 33.74 | 36.98 | 34.56 |
| Arab Heavy–27 | 29.17 | 27.53 | 26.73 | 28.71 | 28.62 | 27.12 | 26.08 | 25.07 | 26.51 | 31.56 | 35.24 | 32.96 |
| Iran Light–34 | 33.85 | 31.06 | 30.10 | 32.03 | 31.82 | 31.20 | 30.17 | 28.04 | 29.25 | 33.51 | 36.78 | 34.49 |
| Iran Heavy–31 | 33.20 | 30.76 | 29.52 | 30.66 | 30.80 | 29.94 | 28.94 | 27.13 | 28.39 | 32.85 | 36.26 | 34.09 |
| Kuwait–31 | 31.59 | 28.65 | 27.63 | 29.93 | 29.80 | 29.04 | 27.97 | 26.45 | 27.89 | 32.58 | 36.04 | 33.64 |
| Iraq Basrah–35 | 33.46 | 30.52 | 29.48 | 31.62 | 31.41 | 30.70 | 29.64 | 27.66 | 29.07 | 33.46 | 36.74 | 34.31 |
| Iraq Kirkuk–36 | 34.38 | 31.41 | 30.38 | 32.46 | 32.18 | 31.44 | 30.35 | 28.17 | 29.65 | 33.90 | 37.15 | 34.61 |
| UAE Murban–39 | 35.84 | 33.08 | 32.09 | 33.82 | 33.51 | 32.92 | 31.89 | 29.29 | 30.50 | 34.42 | 37.53 | 35.16 |
| Nigeria Light–37 | 38.45 | 35.77 | 34.81 | 35.96 | 35.58 | 35.23 | 34.40 | 31.18 | 32.46 | 35.79 | 38.77 | 36.72 |
| Algeria Saharan–44 | 37.89 | 35.30 | 34.27 | 35.50 | 35.04 | 34.79 | 33.76 | 30.66 | 32.40 | 35.03 | 38.05 | 35.83 |
| Libya Zueitina–41 | 37.37 | 34.84 | 33.81 | 35.22 | 34.96 | 34.72 | 33.66 | 30.65 | 32.05 | 34.79 | 37.94 | 35.35 |
| North Sea Ekofisk–42 | 36.09 | 33.70 | 32.77 | 33.60 | 33.33 | 33.30 | 32.77 | 29.99 | 31.40 | 34.58 | 37.95 | 36.11 |

*Based on preliminary estimate of "subsidy" available under U.S. Government crude cost equalization program.

Crude Oil Yield Percentage Patterns in Two Key Refining Areas (*percent*).

| U.S. Gulf Coast—<br>Winter and Summer† | Arab<br>Light | Arab<br>Heavy | Iran<br>Light | Kuwait |
|---|---|---|---|---|
| Premium Gasoline | 6.9 | 5.3 | 7.0 | 6.3 |
| Unleaded Gasoline | 13.7 | 10.5 | 14.1 | 12.5 |
| Regular Gasoline | 25.1 | 19.4 | 25.9 | 23.0 |
| Gas Oil | 25.0 | 0.0 | 28.7 | 7.2 |
| Fuel Oil | 25.7 | 62.3 | 21.2 | 48.4 |
| Percent Sulfur | 2.8% | 2.8% | 2.8% | 2.8% |

| Rotterdam—<br>Winter Basis | Arab<br>Light | Arab<br>Heavy | Iran<br>Light | Iran<br>Heavy | Kuwait |
|---|---|---|---|---|---|
| Naphtha | 6.7 | 5.2 | 6.8 | 6.7 | 6.2 |
| Premium Gasoline | 7.7 | 6.0 | 7.8 | 7.6 | 7.1 |
| Regular Gasoline | 5.8 | 4.5 | 5.9 | 5.7 | 5.3 |
| Gas Oil | 36.6 | 18.5 | 34.3 | 27.7 | 27.0 |
| Fuel Oil | 38.5 | 61.4 | 40.6 | 47.9 | 49.7 |
| Percent Sulfur | 3.18% | 4.0% | 2.46% | 2.52% | 3.95% |
| *Summer Basis* | | | | | |
| Naphtha | 7.8 | 6.1 | 7.9 | 7.4 | 7.0 |
| Premium Gasoline | 8.9 | 7.0 | 9.0 | 8.5 | 8.0 |
| Regular Gasoline | 6.7 | 5.2 | 6.8 | 6.4 | 6.0 |
| Gas Oil | 33.4 | 15.9 | 31.2 | 25.2 | 24.5 |
| Fuel Oil* | 38.4 | 61.4 | 40.6 | 48.0 | 49.7 |

† Refinery with cracking facilities.
*Percent Sulfur same as winter basis.

| Nigeria Light | Algeria Saharan | Libya Zueitina | Tia Juana Light | N. Sea Ekofisk | | |
|---|---|---|---|---|---|---|
| 7.7 | 8.1 | 8.0 | 6.5 | 7.3 | | |
| 15.3 | 16.1 | 15.4 | 12.9 | 14.7 | | |
| 28.1 | 29.6 | 28.2 | 23.7 | 26.9 | | |
| 40.5 | 37.2 | 37.2 | 30.4 | 31.4 | | |
| 6.0 | 5.2 | 8.0 | 23.4 | 16.5 | | |
| 0.58% | 0.58% | 0.67% | 2.8% | 0.59% | | |

| Iraq Basrah | Iraq Kirkuk | UAE Murban | Nigeria Light | Algeria Saharan | Libya Zueitina | N. Sea Ekofisk |
|---|---|---|---|---|---|---|
| 6.5 | 7.9 | 8.1 | 8.9 | 10.7 | 8.1 | 7.0 |
| 7.4 | 9.0 | 9.2 | 10.1 | 12.1 | 10.4 | 8.0 |
| 5.6 | 6.8 | 6.9 | 7.6 | 9.2 | 7.9 | 6.1 |
| 34.6 | 35.8 | 40.4 | 44.8 | 40.2 | 43.6 | 35.5 |
| 41.2 | 35.6 | 30.5 | 23.9 | 22.0 | 23.7 | 38.9 |
| 3.46% | 3.85% | 1.67% | 0.24% | 0.33% | 0.44% | 0.39% |
| | | | | | | |
| 7.5 | 9.0 | 9.3 | 10.0 | 12.1 | 10.5 | 7.9 |
| 8.6 | 10.2 | 10.6 | 11.4 | 13.8 | 11.9 | 9.0 |
| 6.5 | 7.7 | 8.0 | 8.6 | 10.4 | 9.0 | 6.8 |
| 31.6 | 32.4 | 36.7 | 41.3 | 36.0 | 39.0 | 32.5 |
| 40.8 | 35.7 | 30.6 | 24.0 | 21.8 | 24.1 | 39.0 |

# REFERENCES

Adelman, Morris A. 1972. *The World Petroleum Market.* Baltimore: Johns Hopkins University Press.
———. 1979. Unpublished testimony before Judge Andrew Hauk in International Association of Machinists vs. OPEC, No. 78-5012-AAH (C.D. Cal. Aug. 1979).
———. 1980. "The Clumsy Cartel." *The Energy Journal* 1, no. 1 (January): 43.
Alm, Alvin A.; E. William Colglazier; and Barbara Kates-Garnick. 1981. "Coping with Interruptions." In *Energy and Security*, edited by David A. Deese and Joseph S. Nye, pp. 1379-88. Cambridge, Mass.: Ballinger Publishing Company.
Bagehot, Walter. 1873. *Lombard Street.* London: Harry S. King.
Baughman, Martin L., and Paul L. Joskow. 1975. "Energy Consumption and Fuel Choice by Residential and Commercial Consumers in the United States." Energy Laboratory Report MIT-EL75-024, Massachusetts Institute of Technology.
Baughman, M. L., and F. S. Zerhoot. 1975. "Interfuel Substitution in the Consumption of Energy in the United States—Part II: Industrial Sector." Energy Laboratory Report 75-007, Massachusetts Institute of Technology.
Blumenthal, W. Michael. 1979. Speech before the Annual Convention of the Urban League, July 27, 1981. U.S. Treasury Department Press Release, 1981.
Bohi, Douglas R., and Milton Russel. 1978. *Limiting Oil Imports, An Economic History and Analysis.* Baltimore: Johns Hopkins University Press.
Central Intelligence Agency. 1974-81 (various issues). *International Energy Statistical Review.*

Chow, Gregory C. 1960. "Tests of Equality Between Subsets of Coefficients in Two Linear Regressions." *Econometrica* 28 (July): 591-605.
Council of Economic Advisers, Executive Office of the President. 1978. *Economic Report of the President, 1978.* Washington, D.C.: U.S. Government Printing Office.
_____. 1979. *Economic Report of the President, 1979.* Washington, D.C.: U.S. Government Printing Office.
_____. 1980. *Economic Report of the President, 1980.* Washington, D.C.: U.S. Government Printing Office.
Council on Wage and Price Stability. 1980. "Petroleum Prices and the Price Standard." Washington, D.C.: Council on Wage and Price Stability (memo).
Danielson, Albert L. 1979. "The Role of Speculation in the Oil Price Ratchet Process." *Resources and Energy* 2 (Spring): 243-63.
Deese, David A., and Linda B. Miller. 1981. "Western Europe." In *Energy and Security*, edited by David A. Neese and Joseph S. Nye, pp. 181-210. Cambridge, Mass.: Ballinger Publishing Company.
Deese, David A., and Joseph S. Nye, eds. 1981. *Energy and Security.* Cambridge, Mass.: Ballinger Publishing Company.
*The Economist.* 1980 (September 27): 16.
Eckstein, Otto. 1975. "Prepared Statement." In *Oil Price Decontrol*, Hearings Before the Senate Committee on Interior and Insular Affairs, 94th Cong., 1st Sess., p. 340. Washington, D.C.: U.S. Government Printing Office, Serial No. 94-23.
_____. 1978. *The Great Recession with a Postscript on Stagflation.* Amsterdam: North Holland Publishing Company.
The Energy Policy and Conservation Act of 1976, Section 156, Public Law No. 94-163, 89 stat 871 (1975).
Federal Energy Administration. 1976. *1976 National Energy Outlook.* Washington, D.C.: U.S. Government Printing Office.
Feldstein, Martin, and Alan Auerbach. 1976. "Inventory Behavior in Durable Goods Manufacturing: The Target Adjustment Model." *Brookings Papers on Economic Activity No. 2.* Washington, D.C.: The Brookings Institution.
Fisher, Franklin M. 1966. *A Priori Information and Time Series Analysis.* Amsterdam: North Holland Publishing Company.
_____. 1966. *The Identification Problem in Econometrics.* New York: McGraw-Hill.
_____. 1970. "Tests of Equality Between Sets of Coefficients in Two Linear Regressions: An Expository Note." *Econometrica* 38, no. 2 (March): 361-66.
Fisher, Franklin M.; Paul Cootner; and Martin Neil Bailey. 1973. "An Econometric Model of the World Copper Industry." *The Bell Journal of Economics and Management Science* 3, no. 2 (Fall): 568-610.
Fried, Edward R., and Charles Schultz, eds. 1975. *Higher Oil Prices and the World Economy.* Washington, D.C.: The Brookings Institution.

Friedman, Milton R., and Anna Schwartz. 1963. *A Monetary History of the United States*. New York: The National Bureau of Economic Research.

Griffin, James M. 1971. *Capacity Measurement in Petroleum Manufacturing*. Lexington, Mass.: Heath Books.

Hogan, William E. 1980. "Prepared Statement on Oil Taxes and Oil Emergencies." In *Special Oil Taxes*, Hearings Before the Subcommittee on Taxation and Debt Management of the Senate Committee on Finance, 96th Cong., Second Sess., pp. 186-90. Washington, D.C.: U.S. Government Printing Office.

Houthakker, H. S.; Philip K. Verleger, Jr,; and D. P. Sheehan. 1974. "Dynamic Demand Analysis for Gasoline and Residential Electricity." *American Journal of Agricultural Economics*: 56, no. 2: 412-18.

Independent Petroleum Association of America. 1978. "Report of the Supply and Demand Committee." Paper prepared for the annual meeting. Memo.

Jacoby, Henry D., and James L. Paddock. 1980. "Supply Instability and Oil Market Behavior," *Energy Systems and Policy* 3, no. 4 (Fall): 401-23.

Jacobs, Alan. 1981. "Gasoline Markets, Oil Supply Disruptions, and U.S. Policy." Mimeograph, Department of Economics, Massachusetts Institute of Technology.

Kalt, Joseph. 1980. "Federal Regulation of Petroleum Prices: A Case Study in the Theory of Regulation." Unpublished Ph.D. thesis, Department of Economics, University of California, Los Angeles.

Kindleberger, Charles P. 1973. *The World in Depression, 1929-1939*. London: Allen Lane, The Penguin Press.

Krapels, Edward N. 1980. *Oil Crisis Management: Strategic Stockpiling for International Security*. Baltimore: Johns Hopkins University Press.

Lane, Jr., William C. 1980. "Dealing with Oil Supply Disruptions: Meeting Commitments Under the IEA." Unpublished paper presented at the conference on "New Strategies for Managing U.S. Oil Disruptions," Yale University, November 8, 1980.

Lange, David. 1979. "U.S. Industry Facing Still Larger Numbers in 1979." *Oil and Gas Journal* 77 (January 29): 111.

Leviatan, Nissan. 1963. "Consistent Estimation of Distributed Lags." *International Economic Review* 4 (January): 44-52.

Lieberman, Charles. 1980. "Inventory Demand and Cost of Capital Effects." *Review of Economics and Statistics* 62, no. 3 (August): 348-56.

Lovell, Michael. 1961. "Manufacturers' Inventories, Sales Expectations and the Acceleration Principle." *Econometrica* 29 (July): 293-314.

Mampe, Edwin P. 1981. "Prepared Statement." In *Strategic Petroleum Reserve Program*, Hearings Before the Subcommittee on Energy and Mineral Resources of the Senate Committee on Energy and Natural Resources, 97th Cong., First Sess., pp. 113-30. Washington, D.C.: U.S. Government Printing Office.

National Commission on Supplies and Shortages. 1976. *Government and the Nation's Resources.* Washington, D.C.: U.S. Government Printing Office.

National Petroleum Council. 1981. *Emergency Preparedness of Interruption of Petroleum Imports into the United States.* Washington, D.C.: The National Petroleum Council.

*The New York Times.* 1974. (January 18): 1.

Nordhaus, William D. 1973. "The Allocation of Energy Resources." *Brookings Papers on Economic Activity No. 3*, pp. 529-70. Washington, D.C.: The Brookings Institution.

──── . 1980a. "Prepared Statement." In *Special Oil Taxes*, Hearings Before the Subcommittee on Taxation and Debt Management of the Senate Committee on Finance, 96th Cong., Second Sess., pp. 31-33. Washington, D.C.: U.S. Government Printing Office.

──── . 1980b. "Oil and Economic Performance in Industrialized Countries." *Brookings Papers on Economic Activity No. 2*, pp. 241-380. Washington, D.C.: The Brookings Institution.

OECD. 1976-80 (various issues). *Quarterly Oil Statistics.*

*Oil and Gas Journal.* 1973-80 (various issues).

Owens, Charles. 1974. "History of Petroleum Price Controls." In *Historical Working Papers on the Economic Stabilization Program: Part II*, edited by Andrew T. Munroe; Henry H. Perritt; and Johnathon Brock, pp. 1223-1340. Washington, D.C.: U.S. Government Printing Office.

*Petroleum Intelligence Weekly.* 1974-81 (various issues).

Phelps, Charles, and Rodney T. Smith. 1977. *Petroleum Regulation: The False Dilemma of Decontrol.* Los Angeles: The Rand Corporation.

Phillips, L. 1972. "A Dynamic Version of the Linear Expenditure Model." *Review of Economics and Statistics* 54 (November): 450-88.

Pindyck, Robert. 1978. "The Gains to Producers from the Cartelization of Exhaustible Resources." *Review of Economics and Statistics* 60, no. 2 (May): 238-51.

*Platt's Oilgram Price Report* 57, no. 44 (December 19, 1979): 1.

*Platt's Oilgram Price Report.* 1975-81 (various issues).

Plummer, James. 1981. "Policy Implications of Energy Vulnerability." *The Energy Journal* 2, no. 2 (April): 25-36.

Ramsey, J.; R. Rasche; and B. Allen. 1975. "An Analysis of the Private and Commercial Demand for Gasoline." *The Review of Economics and Statistics* no. 57 (November): 502-507.

Roeber, Joseph. 1979. "The Dynamics of the Rotterdam Market." *The Petroleum Economist* 46, no. 2 (February): 49-54.

Rowen, Henry. 1980. "Prepared Statement." In *Special Oil Taxes*, Hearings Before the Subcommittee on Taxation and Debt Management of the Senate Committee on Finance, 96th Cong., Second Sess., pp. 38-40. Washington, D.C.: U.S. Government Printing Office.

Singer, J. Fred. 1981. "The World's Falling Need for Crude Oil." *The Wall Street Journal* (April 21): 20.

Sunley, Emil. 1980. "Testimony." In *Special Oil Taxes*, Hearings Before the Subcommittee on Taxation and Debt Management of the Senate Committee on Finance, 96th Cong., Second Sess., pp. 81-100. Washington, D.C.: U.S. Government Printing Office.

Sweeney, J.L. 1977. "The Demand for Gasoline: A Vintage Capital Model." Draft paper, Department of Engineering-Economic Systems, Stanford University.

Taylor, Carol. 1979. "A Quarterly Domestic Copper Industry Model." *The Review of Economics and Statistics* 61, no. 3 (August): 410-22.

Thomas, Marshall. 1981. "The ABCs of Measuring Oil Market Price Trends." *Petroleum Intelligence Weekly* (February 2): Supplement pp. 2-5.

U.S. Department of Energy. 1979. *Iranian Response Plan: Reducing U.S. Impact on the World Oil Market*. Washington, D.C.: U.S. Government Printing Office.

_____. 1980a. *Final Report to the President on the Oil Supply Shortages of 1979*. Washington, D.C.: U.S. Government Printing Office.

_____. 1980b. *Reducing U.S. Oil Vulnerability, Energy Policy for the 1980's*. Washington, D.C.: U.S. Government Printing Office.

_____. 1974-81 (various issues). *Monthly Energy Review*.

_____. 1973-81 (various issues). *Monthly Petroleum Statement*.

Verleger, Jr., Philip K. 1975. "Price and Income Elasticities of the Demand for Gasoline." Paper presented to the Committee on Energy and the Environment of the National Academy of Sciences. Lexington, Mass.: Data Resources, Inc.

_____. 1979. "The U.S. Petroleum Crisis of 1979," *Brookings Papers on Economic Activity No. 2*, pp. 463-476.

_____. 1981. "Petroleum Price and Allocation Regulations During the 70s: History and Economic Analysis." *Annual Review of Energy* 5.

Verleger, Jr., Philip K., and Dennis P. Sheehan. 1974. "The Demand for Distillate Fuel Oil and Residual Fuel Oil: A Cross-Section Time-Series Study." Paper prepared for the U.S. Council on Environmental Quality. Lexington, Mass.: Data Resources, Inc.

_____. 1976. "A Study of the Demand for Gasoline." In *Econometric Studies of U.S. Energy Policy*, edited by Dale Jorgenson, pp. 177-240. Amsterdam: North Holland Publishing Company.

White House Press Office. 1973a. "The President's Energy Emergency Address." In *The White House Fact Sheet* (November 7).

_____. 1973b. "New Energy Emergency Actions." In *The White House Fact Sheet* (November 28).

_____. 1981. *A Program for Economic Recovery*, Washington, D.C.: U.S. Government Printing Office.

Wright, Brian D., and Jeffrey C. Williams. 1981. "The Roles of Public and Private Storage in Managing Oil Import Disruptions." Working Paper, Department of Economics. Yale University.

# INDEX

Abu Dhabi, 18, 20
"Act of God," 5-6
"Act of war," 5-6
Adelman, Morris A., 15n, 56, 58, 66, 193-194
Ad valorem. *See* Tax(es)
Africa, 46
African light crude oil, 16-20, 61-65, 72-73, 75-79, 85, 87, 201, 203, 264
African producers, response to market changes, 201n
Alaska, 158
Alaskan oil production, 97
Algeria, 18-19
Algerian Saharan crude oil, 264, 271, 273
Allen, B., 126
Allocation (oil), xvii-xviii, xxiv, xxxi
 controls, xix, xxiv, 208-209, 237
 crude supply, 208, 216
 deterrent to stockbuilding, 208-209, 248
 for aviation purposes, 235-236
 IEA plan, 255-256
 mandatory, 208
 product distribution, 208, 216
 through queuing, 232
Alm, Alvin A., 169, 174-175, 249
Alm, Colglazier and Kates-Garnick (study), 169, 174-175, 249

Alternative energy sources, xvii-xviii, xxiii, xxvii.
 *See also* Fuel substitution
American Gas Association, 227n
Arab embargo (1973), xvii, xxiv, xxix, 13, 30-36, 38-39, 41, 46, 58n, 80-81, 89-96, 131, 156, 217, 222, 226, 230
Arab heavy crude oil, 61-65, 271-272
Arab light crude oil, 31, 33-34, 37-38, 57, 61-66, 264, 266-272
 as OPEC marker base, 270
Arab medium crude oil, 264
Arab/Persian Gulf refining center(s), 265-266
Aramco (Arabian American Oil Company), 36
ARCO (Atlantic-Ritchfield Company), 214
Ashland Oil Company, 208n
Australia, 217n
 oil stockpiling program, 218
Austria, 217n
 oil stockpiling program, 218
Auerbach, Alan, 105-107
Aviation fuel, 74, 235-236

Bagehot, Walter, 261
Bailey, Martin Neil (study), 42
Baughman, Martin L., 126
Baughman and Joskow (study), 126

281

# INDEX

Baughman and Zerhoot (study), 126
Belgium, 217n
　oil stockpiling program, 218
Blumenthal, W. Michael, 155
Bohi, Douglas R., 74
Bohi and Russel (stucy), 74
Bradley, Senator Bill, xvii–xviii, 191
Brass River crude oil, 264
British Petroleum (BP), Ltd., 12
Building temperature controls, emergency, 222–226, 237–238, 254–255
　drawbacks of, 224–226
Bureau of Mines, U.S., 91n–92, 94–95

"Cadillac" crude oils, 264
Canada, 217–218
　electricity imports from, 230
Car pooling, 232, 234
Caribbean refining center(s), 265–266
Cartel behavior. See OPEC
Carter Administration
　oil price decontrol, 12n
　emergency fuel conservation measures, 223–226, 228, 232–234
Carter, President Jimmy, 164, 191, 208n, 221, 233
Central Intelligence Agency (CIA), 34, 41, 80–81, 83–84, 96
Chow, Gregory C., 61, 64
Cities Service Company, 214
Civil Aeronautics Board (CAB), 235–236
Coal, increased use of, as emergency fuel substitute, 222, 226–228, 230, 238, 254
Colglazier, William, 169, 174–175, 249
Commodity market(s)
　importance of, 30
　OPEC price adjustments and, 4, 20, 55–66, 123
　price increases on, during disruption, 4–5, 13–14, 20–21, 29–53, 120
　products traded on, 11
　supply disruption and, 10–14
　See also Rotterdam, Spot market(s)
Conoco (Continental Oil Company), 214
Conservation, fuel
　cuts in consumption, 4, 112, 221–244, 253, 260

emergency measures, xviii, xxiv, xxvii, xxix–xxx, 221–244, 246, 253–255, 260
　minimization of effects of, 5
　personal, 223
　price induced, 53, 137, 160, 167
　standby programs, 233, 244
Consumer(s)
　benefits to, xxvii, 156
　contribution to increasing prices, xxvi
　control of stock withdrawal, xxix, 246–253
　efforts of collective action by, 221–222
Consumer demand, xxiv, xxviii, 120, 143–145, 179, 245
Consumer inventories, 4–5, 22–24, 26, 186, 215–216
Consumer prices
　behavior of, 67–87, 123–146, 156–158, 165, 167–168, 211
　European, 80–87
　U.S. behavior during Iranian crisis, 75–79
Consuming countries
　government encouragement to stockpile, 199
　response to disruption, 179, 199, 255–256
　transfer of wealth from, xxviii–xxix, xxxi, 245, 251
Consumption
　cuts in, 4, 112, 221–244, 253, 260
　emergency reduction policies, 221–244
　necessity of restricting, xxiv, xxvii–xxxi, 179–180
　postponement of, 18–181
　priority users, xxiv, 216
　See also Conservation
Contract market, 40
Contract prices, 75, 77, 83–85, 87, 111
Control(s). See Allocation, Price controls
Cootner, Paul (study), 42
Cost of Living Council, 74
Cost pass-through system, 156, 211
Council of Economic Advisers, 12n, 56, 74, 165, 210
　The Economic Report of the President, 109

INDEX 283

Council on Wage and Price Stability (COWPS), 73-75
*Crude Petroleum, Petroleum Products and Natural Gas Liquids*, 92, 94-95

Danielson, Albert L., 47n, 91, 123, 149
Deese, David A., 26, 215, 217, 224n, 249, 258, 260
Deese and Nye (study), 26, 249, 258, 260
Demand curve, 13-15, 21-25, 43-45
"Demand restraint," 256
Denmark, 217n
  oil stockpiling program, 218
Department of Energy (DOE), U.S., 35, 89, 91n, 95, 98, 100, 103, 122, 148, 189, 208, 222-223, 225-229, 234, 237-238, 249
  regulations, 74-75
  *Monthly Energy Review*, 76, 98, 109
  *Monthly Petroleum Statement*, 98-99
  *See also* Iranian Response Plan
Department of the Interior, U.S., 92, 94
Diesel fuel, 67, 74-77
  emergency conservation measures, 232-233
  European prices of, 80-84
  price increases of, 75-77
  shortages of, 96
*Direct Federal Action on Oil Imports: An Analysis of Import Fees and Quotas* (CBO study), 251
Discount rates, 8n, 10
Distillates, 67, 223, 266
"Distress sales," crude oil, 264
Domestic crude production, 74, 91n-92, 94-95, 97-99, 102-103
Driver education programs, 232, 234
Dubai crude oil, 264

Eckstein, Otto, 56, 164
Economic chaos, xxiii, xxi, 4
  *See also* Panic buying and selling, Transfer of wealth
Economic Stabilization Act (ESA), 122
*Economist, The*, 3-5, 89, 122
Egypt, 33

Electricity, emergency conservation measures, 223, 229-231, 238-239
Electricity wheeling
  defined, 228
  as emergency conservation measure, 223, 228-229, 230n, 238, 254
Embargo(es), 9-10, 13, 208n
  1979 U.S. embargo on Iran, 37
  *See also* Arab embargo (1973)
"Emergency oil drawdown obligations" (ERDO), 256-257, 260
Emergency Petroleum Allocation Act (EPAA), 73
*Emergency Preparedness of Interruption of Petroleum Imports into the United States* (NPC), 218, 222, 257
End user allocations of heating oil, 215, 223
End user markets, 264
Energy Policy and Conservation Act (EPCA), 213, 223, 233
Entitlements system, 31n, 36-37, 74, 80, 111, 167
*Erdoelbevorrantungsverband* (EBV), 211
Europe (Western), 67, 80-82, 94-95, 101-102, 105, 115-119, 121-122, 188, 222, 224
European Economic Community (EEC), 84
  relevance of spot market to, 45
Exxon Corporation, 12, 67, 213-214

Federal Aviation Agency (FAA), 236n
Federal Energy Administration (FEA) (study), 126
Federal Power Act, 228-229
Federal Reserve, 188
Feldstein, Martin, 105-107
Feldstein/Auerbach specification, 107
*Final Report to the President on the Oil Supply Shortage of 1979* (DOE), 89
Fisher, Franklin M. (study), 29, 42, 43n, 61, 64
Fisher, Cootner and Bailey (study), 42
Four-day work week, 232, 234
France, 80, 258
  consumer prices in, 80-84
  emergency energy programs, 223-224, 231

major consumer, 258
oil stockpiling program, 120, 199, 211, 217-218
price controls in, 157, 211
Fried, Edward R., 165, 189-190
Fried and Schultz (study), 165, 189
Friedman, Milton R., 15
Friedman and Schwartz (study), 15
Fuel substitution/switching, 222-224, 226-230, 238, 254, 256
Fuel Use Act, 227
Future income, 7-8
Future prices, 4, 8n, 9n, 10, 66, 180-181

Gas. *See* Natural gas
Gas oil, 11, 13, 84-85, 223, 266
Gasoline, 67, 216, 266
  emergency conservation measures, 89, 231-234
  European prices of, 80-84
  premium, 11, 266
  price increases of, 75-77
  regular, 11, 13, 85, 266
  tax on, xxix, 169-170
  unleaded, 266
Gasoline rationing, 89, 93, 97, 222-223, 231-232, 236-237, 239
  difficulties of, 237
Great Britain, 217n. *See also* United Kingdom
Greece, 217n
  oil stockpiling program, 218
Griffin, James M. (study), 107, 125
Gross margin limitations, 156
Gross product worth (GPW) of refined crude oil, 267-268
Gulf Coast, U.S.
  commodity market(s), 29, 111, 264-265
  refinery, 75n
Gulf Oil Company, 55-56, 209. 214

Heating oil, 67, 74, 104, 166
  emergency conservation of, 223
  price increases of, 75-78, 78n
  shortages of, 97
Hogan, Professor William, 167
Houthakker, H. S., 126
Houthakker, Verleger and Sheehan (study), 126

Income tax. *See* Tax(es)
Independent Petroleum Association of America (IPAA), 97, 100
Indonesian Minas crude oil, 271
"Industrial petroleum reserve," 213
"Inland product realizations," 84-85
"Insurance" stocks, 215
Interest subsidies, 204, 206-207, 215, 247-248
Internal Revenue Service, U.S., 170
International Energy Agency (IEA), xxiv-xxxi, 38, 217, 255-260
International Energy Program (IEP), 256
*International Energy Statistical Review* (CIA), 80-84
*International Oil Developments*, 96
Interstate Commerce Commission, 235
Inventory accumulation
  crude in transit, 90-91
  during disruptions, 89-122
  effects of abnormal weather on consumption, 91, 120
  imported crude oil, 90-95, 99, 102
  Lovell model of inventory behavior, 105-107
  petroleum stockholding locations, 90
  reasons for, 104-111
  refined products, 90, 92-95, 97, 102
  relationship between stocks and profits, 111-114
  seasonal influence on, 49, 91
  *See also* Speculative inventories, Stockbuilding, Stockholdings
Inventory drawdown, emergency
  immediate, 127, 246, 249, 254
  importance of, 246-247, 249
  of government stocks, 246, 249
  of private stocks, 190, 246, 249
  regulated, 247, 249
  through market mechanisms, 247, 249-250
  to moderate price increases, 172, 179-181, 186, 247, 249
Inventory levels
  critical determinant of price behavior, 146
  response to prices, 159-163, 172, 199, 246, 249-250

INDEX 285

Iran, 6, 10-11, 33-34, 208n, 158
  oil workers strike in, 6, 33-34, 42
  Shah of, 6, 33-34, 194
Iranian crisis (1979), xvii, 30-34,
    37-39, 41, 75-77, 80-82, 89-91,
    96-97, 99-102, 120, 122, 134,
    147, 156, 202, 209, 221-223,
    226, 228, 259n
  interruption of supply, xxiv, 12,
    33-34, 36-37, 45, 96, 101, 208
Iranian heavy crude oil, 37, 264,
    271-272
Iranian light crude oil, 37, 264,
    271-272
Iranian Response Plan (DOE), 226,
    228-230
Iran/Iraq War (1980), 31-32, 34,
    36-39, 42, 58n, 134, 139,
    149-150
  disruption of supply due to, xvii, 3,
    10-11, 114
Iraqi Basrah light crude oil, 264, 271,
    273
Iraqi Kirkuk crude oil, 271, 273
Ireland, 217n
Israel, 33
Italy, 217n
  consumer prices in, 82-84
  oil stockpiling program, 218

Jacobs, Alan, 237, 277
Jacoby, Henry D., 58
Jacoby and Paddock (study), 58
Japan, 94-96, 115-119, 121-122,
    217n
  commodity market, 264-265
  confrontation with U.S. over
    embargoed Iranian crude stocks, 37
  consumer prices in, 80-82
  oil stockpiling program, 101, 199,
    218
  price controls in, 157
  refined products supplied to,
    101-103
  response to Iranian crisis, 12
Jet fuel, 11
Joint tariff conservation measures,
    242-244
Joskow, Paul L., 126

Kalt, Joseph, 74, 80, 82, 156
Kassenbaum, Senator, 213

Kates-Garnick, Barbara, 169,
    174-175, 249
Kerosene, 67
Kharg Island loading facility strike, 12
Krapels, Edward N., 258
Kuwait, 18-20, 38, 158
Kuwait crude oil, 57, 264, 271-272

Lane, William C., Jr., 204
Leviatan, Nissan, 60n
Libya, 18-19
Libyan Zueitina crude oil, 264, 271,
    273
Lieberman, Charles, 105
Los Angeles, electricity conservation
    measures in, 230-231, 238
Lovell, Michael, 105-107, 114
Luxembourg, 217n
  oil stockpiling program, 218

Mampe, Edwin P., 204
Marathon Oil Company, 213-214
Marginal cost curve, 70
Mideast 31-gravity crude oil, 264
Mideast 34-gravity crude oil, 72-73,
    264
Motor Carrier Act of 1980, 235
Multinational oil companies, 10, 21

Naptha, prices of, 84-85, 265-267
National Commission on Supplies and
    Shortages, 217, 219-220
National Energy Plan, 12, 193n
National Petroleum Council (NPC),
    158n, 174n, 217-218, 222,
    225-232, 234, 236, 238, 254,
    256-258.
    See also Emergency Preparedness . . .
Natural gas, increased use of, 91n, 92,
    94-95, 98-99, 222-224, 226-
    228, 230, 238, 254
"Netbacks," 15n, 16, 20, 55, 57,
    139n. See also Spot values
Netherlands
  emergency conservation (gas)
    program, 231
  1973 OPEC embargo and, 9, 13
  oil stockpiling program, 217-218
New York Times, The, 93, 221
New Zealand, 217-218

Nigeria, 18-19
  1979 embargo against U.K., 10
Nigerian crude oil, 12, 78, 80
  Bonny light crude oil, 57, 75, 264, 271, 273
Nixon Administration, 223
  emergency conservation plan, 223
  emergency consumption regulation, 223, 225-226, 232-234
Nixon, President Richard M., 233, 235
Nordhaus, William D., 40-42, 58, 60, 165, 171, 174, 199, 210
North Sea Ekofisk crude oil, 264, 271, 273
Norway, 217
Nye, Joseph S., 11, 26, 249, 258, 260

"Odd-even" plans for fuel conservation, 232, 234
*Oil and Gas Journal*, 97, 100, 114
Opportunity cost, 109-110
Optimal intertemporal producing plans, 7
Organization for Economic Cooperation and Development (OECD), 23n, 33-34, 90-91, 101, 103n, 108, 115, 122, 148, 157-159, 168-169, 171-172, 190, 193, 195
  *Quarterly Oil Statistics*, 102-103, 108
Organization of Arab Petroleum Exporting Countries (OAPEC), 91
  1973 embargo, 9, 33
Organization of Petroleum Exporting Countries (OPEC), xxvi-xxvii, 18-19, 258
  cartel behavior, 20, 56, 66, 245
  1973 embargo, 9-10
  meetings, 47n, 55, 66
  price induced production cuts, 26, 163
  pricing adjustment, 3-5, 20, 40, 55-67, 135-137, 158-159, 159n, 160, 175, 187-188, 245, 253
  response to emergency tariffs, 187-188, 193-197
  response to aggressive pricing, 163-164
  unstable production levels, 123
Owens, Charles, 156

Paddock, James L., 58
Panic buying, 26, 66, 249, 260
Panic selling, 66
Persian Gulf, 46
  refining centers in, 265-266
Personal vehicle use, emergency reduction of, 231-234, 238
*Petroleum Intelligence Weekly (PIW)*, xx, 12-13, 16-17, 19, 30, 34-37, 61-62, 71, 73, 75, 81, 83-85, 135n, 148, 201, 221, 259, 263-268
Phelps, Charles, 80, 156
Phillips, L. (study), 126
Phillips Petroleum Company, 214
Pindyck, Robert (study), 125
*Platt's Oilgram Price Report*, 67
*Platt's Oil Price Service Organization*, 13, 32, 34, 55-57, 61-62
Plummer, James, 189-190, 193-194
P.O.E., 34
Postings, crude oil price, 4, 58n, 170
Power pool(ing), 238
  defined, 229
Power wheeling, 227, 229, 237-239
"Precautionary" stocks, xxx, 199, 202, 215-216.
  *See also* Speculative inventories
Price controls, 155-167
  cost based, 157-158
  deterrent to stockbuilding, 200, 207-209, 215, 219
  effectiveness of, 80, 156-165
  effect on consumer prices, 80, 156-158, 165, 167-168, 211
  federal, xvii-xviii, xix, xxiv, 73-87
  gross margin limitations, 156
  ineffectiveness of, xvii-xviii, xxv, 80, 155-165, 248
  mandatory, xxiv, 159
  product plus crude oil control, 158-159, 161-163
  product price control, 73, 158-159, 161-162
  threat of, xviii
  U.S. program, 155-165
  voluntary, xxiv
Price setting, aggressive, 158, 160-165, 215
Price supports, 210, 248
Priority users, xxiv, 216

INDEX 287

Producing nations, transfer of wealth to, xxviii–xxix, xxxi, 245, 251
Profit ceilings, 156
*See also* Price controls

Qatar, 18, 20
Quota/auction. *See* Tax(es)

Ramsey, J., 126
Ramsey, Rasche and Allen (study), 126
Rasche, R., 126
"Ratcheting," oil price, xxvii, 123, 151
Rationing, xvii. *See also* Gasoline rationing
Reagan, President Ronald, 156, 191, 224
Recreational uses of fuel, emergency restrictions on, 232–234
Refining centers, 84, 265
Residual fuel oil, 11, 13, 126
 price increases of, 75–78, 78n, 84–85
Roeber, Joseph, 30
Rotterdam market, xxv, 269
 oil price setting by, 29–30, 263–264
 price increases on, 4, 12, 33
 product prices, 13, 61, 75, 78, 80, 84
 refining center, 265, 267–268
 *See also* Commodity market(s)
Rowen, Henry, 169n
Russel, Milton, 74

Saudi Arabia, 18–20, 158
 cuts in production, 34–37, 42, 46, 58n, 147
 increased supply during Iran/Iraq War, 10, 36–38, 42, 148, 151
 monopolistic control of production and price, 6n, 55
 response to disruption tariffs, 193
Schlesinger, James, 12, 193n, 221
School bus utilization, emergency, 232, 234
Schultz, Charles, 165, 189–190
Schwartz, Anna, 15
Seasonal fluctuations in demand, 33, 49, 91, 207

Sheehan, Dennis P., 126, 190
Shell Oil Company, 214
Shipping lags, 120
Singapore
 commodity markets in, 29
 export refining center, 265–266
Smith, Adam, 29
Smith, Rodney T., 80, 156
Social Security, 191–192, 252
Sohio (Standard Oil Company of Ohio), 214
Spain, 217n
 oil stockpiling program, 218
*Special Oil Taxes*, 167, 199
Speculation, xxvii–xxviii
 government encouragement of, 199–200, 211–213
 potential benefit of, xxix, 172, 199
 profits from, 172, 247–248
Speculative inventories, xxvi–xxvii, 4, 49, 199–220
 advantages of, 199
 building of, 4–5, 89, 199
 deterrents to, 199–200, 207–208, 215, 219
 incentives to build, 167–172, 199–207, 209–213, 215–216, 244–248, 255, 260
 incentives to sell, 172, 181, 190, 210, 246–247, 249, 254, 258, 260
 liquidation of, 89, 196
 storages costs, 248
 supplier, 200–216
Speed limit enforcement, 232, 234–235
Spot market(s), xxx, 11–14, 20–26, 29–87
 *See also* Commodity market(s), Rotterdam
Spot values (market prices)
 pattern of, 263–266
 world cuts in consumption to moderate increases of, 240–242
Standard Oil Company of California, 214
"Standby" stocks, 215
Stanford University, 169n
"Sticker" plans for fuel conservation, 232, 234
Stockbuilding (inventory), xvii, xxvii, 199–220

by consumers, 4-5, 22-24, 26, 215-216
by oil companies, 3-5, 26, 122, 200-209, 213, 216
deterrents to, 199-200, 207-208, 216
during times of shortage, 67, 89, 210, 215, 245, 259-260
during times of surplus, 172, 202, 204, 210, 246-247, 253
government regulated, 199, 211-213, 219
impact of political conditions on, 248
private, xxvii, 4-5, 199-200, 217
public, 217, 219
rate of acquisition, 120-121, 160n
speculative, 205, 247-248
strategic, 213, 249-250
Stockholdings (inventories), 199-220
cost of, 110
government owned, 199, 219, 249
precautionary, xxx, 199, 202, 215-216
privately owned, xxvii, xxix, 216, 219-220, 246, 249
publicly owned, xxix, 217, 219-220, 246
speculative, 202, 204-205
standby, 215
strategic, xxvii, xxix.
*See also* Strategic petroleum reserves
support programs for, 210
Stock levels, to moderate price increases during disruptions, 52-53, 159-163, 172, 199, 246, 249-250
Stock management, 89, 100, 122, 246-250, 253
Storage cost(s), 105, 107, 110
of speculative stocks, 248
subsidized, 248, 250
Storage programs, 211-212
Strategic petroleum reserves, xvii, xxvii, 101n, 213, 249, 250
Strategic Petroleum Reserve Plan (SPRP), 249-250
Subsidization, 80, 214-216
interest rate, 204, 206-207, 215, 247-248

Sunday driving, emergency ban on, xxx, 232, 247-248
Supply curve, 6n, 7-11, 21-22, 24-25, 43, 68, 70, 71n
Sweden, 217n
oil stockpiling program, 218
Switzerland, 217n
oil stockpiling program, 218
Syria, 33

Tariff(s), xviii, xxvii, xxix, 167-197, 247
ad valorem, 168-169
alternative levels, 179, 183-188
constant, 179, 182
disruption, xxviii-xxix, 168, 178, 188-190, 193-194, 209, 215-216
economic consequences of, 188
effectiveness of proposals, 179-183, 196-197
effects of, on oil markets, 168-169
graduated, xxviii, 196n, 246, 254
on imported crude oil and products, xxvii, 168-169, 171-173, 175-178, 190, 209, 215, 246-247
regular, 193
recycled, xxviii, 187-188, 190-193, 197, 246, 251-253
specific per barrel levy, 168, 170, 173-175
standby, 209, 216, 246-247, 253
temporary, 194
to discourage consumption, 247
to moderate price increases, 168, 172, 190
variable, 178-184, 187, 196-197
*See also* Joint tariff conservation measures
Tax(es), 167-197, 200
constant, 168
consumer, xxix, 81, 168, 209, 247
corporate profits, 252
effectiveness of, 168, 196-197
effects of, on oil markets, 168
excise, 170-171, 189, 196
gasoline, 169-170
general reduction of, 191-192
income, 191-193, 202, 252
quota/auction, 168, 171-175, 195-196
rebates, 169, 191-192, 246

recycled, 190–192, 251–253
refiner(y), xxix, 168–170, 172–176, 178, 196, 209, 247
severance, 168, 170, 172–174
standby, xxix, 216, 247
to moderate consumer prices, 168, 172
value added (VAT), 181, 188, 192–193, 252n
variable, 181
windfall profits, 168–175, 188, 196, 247, 252
Tax credits, 204
Tax exemptions, 206
Tax legislation, 190–191
Tax liability, 191, 204
Tax rebates, 169, 191–192, 246
Tax subsidies
on stockholdings, xxvii, 204–207, 213, 247–248, 253
on storage costs, 204, 248, 250
Tax withholdings, 191–192, 252
Taylor, Carol (study), 42
Thomas, Marshal, 263–273
Trade Expansion Act of 1962, 190
Transfer of wealth
within the United States, xviii
from consuming to producing countries, xxviii–xxix, xxxi, 245, 251
Transportation fuel, emergency conservation of, 232–236
Turkey, 217n

Union Oil Company, 214
United Arab Emirates (UAE) Murban crude oil, 271, 273
United Kingdom, 181
consumer prices in, 80–84
oil stockpiling program, 218
1979 Nigerian embargo against, 10
United States, oil stockpiling program, 217–218
U.S. Congress, 156, 164, 191–192, 193n, 206, 232–233
U.S. Congressional Budget Office, 187n, 189–190, 251
U.S. crude oil buy/sell program, 120

U.S. House of Representatives, 233, 237
U.S. National Income and Products Accounts (Dept. of Commerce), 108
U.S. Secretary of Energy, 213, 248, 260
  *See also* Schlesinger, James
U.S. Secretary of the Treasury, 12n
U.S. Senate, 233, 237
  Committee on Finance, 167, 199
  Interior Committee, 164
U.S. State Department, 37
U.S. Treasury Department, 252
U.S. Virgin Islands, refiners in, 97n

Vehicle inspection, 232, 234
Venezuelan Tia Juana crude oil, 271, 273
Verleger, Philip K., Jr., 126, 190
Volumetric allocatioans, 248
"Voluntary conservation," 242

*Wall Street Journal, The,* 55
Weather, and demand relationship, 115, 120, 127
Weekend closing of filling stations, emergency, 232–234, 239, 254, 260
West Germany, 80, 182
  consumer prices in, 80–86
  emergency energy programs, 223–224, 231
  oil stockpiling program, 120–121, 199, 211–212, 215, 217–218
White House Press Office releases, 223, 225–226, 228–229, 234, 238, 251n
Williams, Jeffrey C., 219–220
Worldscale, 268
Wright, Brian D., 219–220
Wright/Williams study, 219–220

Yamani, Sheik, 55, 163n
Yield patterns, crude oil, 266–267

Zerhoot, F.S., 126

# ABOUT THE AUTHOR

**Philip K. Verleger, Jr.** holds a Ph.D. in economics from the Massachusetts Institute of Technology. He was a lecturer and senior research scholar at the Yale School of Organization and Management when the draft of this book was completed. Prior to joining the faculty at Yale, he had served as a special assistant (for energy matters) to the assistant secretary for economic policy at the U.S. Treasury; as a senior staff economist for the president's council of economic advisers; as director and developer of the energy service at Data Resources Inc; and on the faculty of the University of California, Santa Barbara. He is currently a lecturer at Yale and a principal with the firm of Booz Allen & Hamilton.

Mr. Verleger has served as a member of the Energy Policy Task Force which advised President Ronald Reagan during the election campaign in 1980, as an adviser to the National Petroleum Council's study on emergency preparedness, as an editor of the *Energy Journal*, and member of the Scientists' Institute for Public Information panel on emergency petroleum storage.